BODIE'S GO

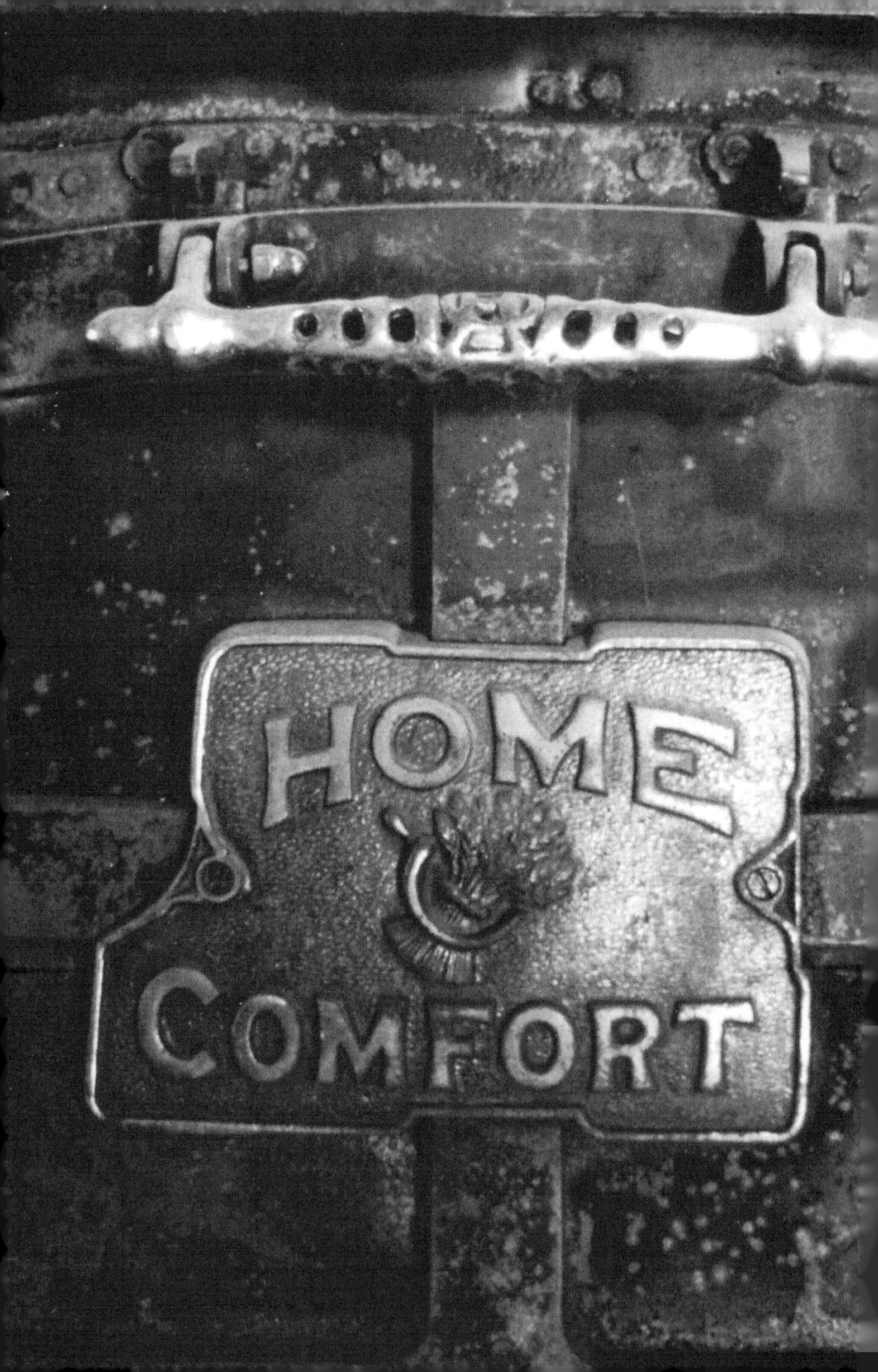
HOME
COMFORT

MARGUERITE SPRAGUE

Bodie's Gold

TALL TALES

AND TRUE

HISTORY

FROM A

CALIFORNIA

MINING TOWN

UNIVERSITY OF NEVADA PRESS ▲▲ RENO & LAS VEGAS

University of Nevada Press, Reno, Nevada 89557 USA

www.unpress.nevada.edu

Copyright © 2003 by University of Nevada Press

Photographs copyright © 2003 by Marguerite Sprague, unless otherwise noted

All rights reserved

Manufactured in the United States of America

Library of Congress Cataloging-in-Publication Data

Sprague, Marguerite, 1958–

Bodie's gold : tall tales and true history from a California mining

town / Marguerite Sprague.

p.   cm.

Includes bibliographical references and index.

ISBN: 978-0-87417-856-2 (pbk. : alk. paper)

1. Bodie (Calif.)—Gold discoveries. 2. Bodie (Calif.)—History. 3. Frontier and pioneer life—

California—Bodie. 4. Gold miners—California—Bodie—History. 5. Mining camps—

California—History. 6. Ghost towns—California. 7. Bodie (Calif.)—Biography.

8. Interviews—California. 9. Folklore—California—Bodie. I. Title.

F869.B65—S68   2003

979.4'48—dc21   2002151291

The paper used in this book meets the requirements of American National

Standard for Information Sciences—Permanence of Paper for Printed Library

Materials, ANSI/NISO Z39.48-1992 (R2002). Binding materials were selected for

strength and durability.

University of Nevada Press Paperback Edition, 2011

This book has been reproduced as a digital reprint

Frontispiece: Home Comfort Stove

For Robert, Nettie, and Lilah: with you, Eureka!

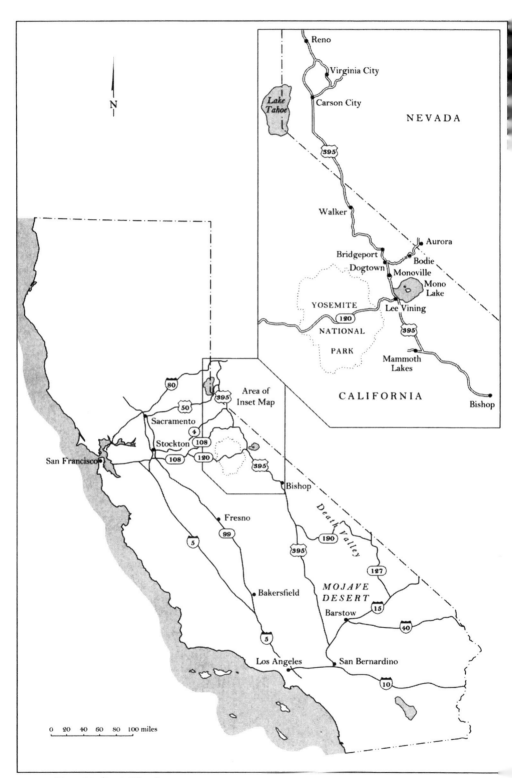

Map of Bodie, California

CONTENTS

ILLUSTRATIONS

PREFACE

Tales of my wild Uncle Marion, many told by Marion himself, enlivened my relatively ordinary childhood. A visit to his Sierra home was like a visit to a fabulous adventure book: full of scary fun, but with a guaranteed happy ending. I still recall him rousting me out of my warm sleep at 4 A.M. to plop me into his old Army surplus jeep—no airbags, no seatbelts, no *doors*. Off we went, careening through the frozen countryside to his favorite patch of Rock Creek, to teach me proper trout fishing. After a trout breakfast, he took everyone up to Bodie, his former home. That day we weren't allowed to go into the town (the state park had not officially opened yet), and had to content ourselves with a view from Bodie Bluff.

Years later, to pay homage of sorts to my rowdy and influential (on me, anyway) uncle, I visited Bodie again. Uncle Marion was gone, but I certainly felt closer to him at his old house in the ghost town than at his grave in a southern California city cemetery.

Bodie gnawed at me. What began as a memoir for the family became a photography project as well. Then, being a technical writer, I had to make sure the "facts" I found were indeed true. I found just enough falsehoods to make me keep digging. You could write a Bodie encyclopedia, if you had the time. The amount of information (much of it of dubious accuracy) is staggering. I had to draw the line somewhere to keep the size reasonable. The result is this book.

Several groups of Bodieites have been either ignored or trivialized in the past, and they deserve more serious consideration. The two most numerous groups are the Kuzedika, the American Indians who lived in the area before the European-Americans showed up, and the Chinese. Another group is the people who lived in Bodie after the boom subsided. I have tried to describe some of these lesser-known Bodieites and their presumed perspectives on Bodie life. For each group, I have drawn on different resources.

Although the Kuzedika are a recognized tribe, and many Kuzedika people are alive today, there is not much published information about the tribe specifically. They are grouped with the Paiute and/or Shoshone people, whose original territories are to the north, south, and east of Mono Lake. Information about the people who lived in the immediate Mono Lake area is harder to come by. Many of them moved around the eastern Sierra seasonally, following the food supply, making them even more difficult to pinpoint geographically. I have relied heavily here on information from recorded interviews conducted in Bridgeport by anthropologists in the early 1930s. The quotes from Kuzedika people are from these interviews. The interviewees were Kuzedika in their 80s and 90s, people who grew up before and during the boom time of the Bodie area. The interviews can be found in the Ethnological Collection of the Bancroft Library, University of California at Berkeley. Some contemporary Kuzedika voiced the strong desire not to have details of their private rituals appear in popular publications; therefore I have not included such information in this book.

I found in talking with a number of American Indians (Kuzedika, Owens Valley Paiute, Pomo, and other nations) that there are no universally accepted terms to use when you are talking about the people who were living here before the European-American settlement movement of the eighteenth and nineteenth centuries. Some prefer the term "Native American"; more seem to prefer "American Indian." Still others point out that all those terms are Anglo inventions and are therefore irrelevant to their identities. I am not taking on the argument; everyone seems to have a good point. However, I must use some term, and so I have used "American Indian" to refer to all the people living in what is now the continental United States before the European-American settlements began. I use the term "Kuzedika" to refer to the people who were living in the Mono County area when W. S. Bodie first arrived; it is their own name for themselves and is therefore indisputably correct. When I quote from or refer directly to newspaper items from that time, I use the term "Paiute" where it was the term used in the newspaper.

For information about Chinese Bodieites, I have relied on Bodie newspaper articles from that time, books about the Chinese (and other Asians) in California during the Gold Rush, and conversations with researchers who have been examining the lives and times of the Chinese in California. Interestingly, none of the researchers I spoke with had ever heard of Bodie.

I should also note that when describing Bodie society in its late nineteenth-century heyday, I use the term "proper" to refer to people, behavior, and so on that corresponded to the Victorian ideal of correct and desirable. This invariably meant Anglo-Saxon, East Coast/European in manner and style, and unsullied by rowdy activities that strike many of us today as rather fun. My use of this term by no means implies a judgment on my own part: I personally consider "proper" to be a highly variable property, dependent upon specific situations and generally irrelevant when one is speaking of humans in their varying forms and ways.

Quotes in this book come directly from their noted sources. Therefore, if there is some deviation from technically correct terminology (such as "the Bodie and Benton Railway" versus "the Bodie and Benton Railway and Commercial Company" or "Mono Lake Railway and Lumber Company"), that is because the original sources used the quoted words.

Although Bodie's ambiance easily conjures up visions of life here in the late 1800s, many of the items on view in the town today actually date from Bodie's later days. Some latter-day Bodieites have sent notes to the park over the years, telling of incidents and happenings in the town long and not so long ago, and some were even interviewed by park personnel. I had the distinct pleasure of interviewing eight former Bodieites at different times, one of whom was my own Uncle Jack Robson. I am grateful to them for their time and hospitality. I would like to thank them all: Mr. Bob Bell, Mr. and Mrs. Gordon Bell, Mr. and Mrs. Ed Goodwin, Mrs. Arena Bell Lewis, Mrs. Marilyn Fern Gray Tracy, Mr. and Mrs. Voss, Dr. Jack Robson, and Mrs. Lauretta Gray, may she rest in peace. All of them were enormous fun to talk with. Bodieites are a very hospitable and charming crowd, not at all the rude gunslingers of reputation!

Thanks to Bill Griffith, creator of *Zippy the Pinhead,* for his generous permission to include Zippy and his unique perspective on Bodie.

The historical photographs in this book come from private collections, the State of California Photographic Archives, and the Bancroft Library at the University of California at Berkeley. Special thanks to Marjorie Dolan Voss, Fern Gray Tracy, and Gordon and Jeanne Bell for their generosity in sharing their photos. Thanks to Wil Jorae at the California State Photographic Archives for his tireless assistance. Thanks to Timothy and Judy Hearsum for their assistance and moral support; additional thanks to Timothy for his photographic genius and beautiful portraits of some of the interviewees included herein.

I appreciate very much the time and assistance of Mr. Joe Lent of Bridgeport and the staff of the Bancroft Library at the University of California, Berkeley. I am grateful to Dr. Judy Yung of the University of California at Santa Cruz for taking time out of a very busy schedule to help this total stranger.

I appreciate the expertise and advice of several former employees of Bodie Consolidated and Galactic Mining Corporations, including Bruce

Gandosi, Horngdie Lee, and Mark Whitehead, in describing the geology of Bodie. Thanks to my father, Dr. Robert W. Sprague, for his many consultations on the relevant chemistry as well as anecdotes from Bodieites he has known, and his undying support and encouragement. Thanks also to my stepmom, Florence Sprague, for her consultations as my oracle of the English language: Ain't none better!

And enormous thanks to the past and present Bodie State Historic Park staff for their assistance and enthusiasm, particularly Julia Hayen, "Bodie Jack" Shipley, Mark Pupich, Dydia Delyser, and Brad Sturdivant. Thanks also to California Department of Parks and Recreation staff members Bill Lindeman and Noah Tilghman for their help.

Deepest thanks to my family for their support, encouragement, and patience throughout this project.

And finally, thanks to my Uncle Marion, may he rest in peace.

BODIE'S GOLD

Doorway interior.

You could no more have another Bodie in these times
than you could have another Trojan War with the
present-day Greeks as participants. The comparison
of the California pioneers with Homer's heroes is not
far fetched. They had much in common.

— G R A N T   H .   S M I T H (1925)

Welcome to Bodie

"Welcome to Bodie!" is the first thing you hear when you arrive, if, like most visitors, you arrive during the warm season when the entrance is staffed. From the park gates, Bodie sprawls across the shallow desert bowl, surprisingly quiet after all these years. This is the baddest town from the bawdy Wild West? This is the dangerous hamlet that spawned tales of the Bad Man from Bodie? Ah yes, it is.

Looking at it today, it's difficult to imagine Bodie ten times bigger, with 10,000 Bodieites coming and going, and mines and mills running twenty-four hours a day. But that's how it was in the boom time. And then, as quickly as it had exploded, Bodie died. Almost. The population dwindled away until there were just a few hundred left. Finally, during World War II, the government called a halt to all gold and silver mining. Bodie's mines closed down, and the last of the hardy Bodieites headed out. Now it's a state park, preserved but not restored, sitting in mute testimony to Bodie's rowdy days and tough-as-iron citizens.

Bodie sits as Bodie was left. There are no gussied-up storefronts, no actors in cowboy duds, no player pianos tinkling out atmosphere. When you come to Bodie, you peer through windows, forced to keep a respectful distance from what was once a big, booming, dangerous gold town of the western frontier. You explore Bodie the same way you paw through Grandpa's dresser drawer: looking for another time, another life, and trying to step into it, to know what it was like to be *them,* to be *there.*

Main Street, Bodie.

Bodie today is so much a composite of fantastic tales of riches and ruin, spent bullets, spent lives, broken glass, and splintering boards that it is hard to separate fact from fantasy. The town's history is fairly murky: On close scrutiny, even the most solid-seeming facts become subject to doubt. We therefore rely on a combination of facts documented in records of the time, on information that is the consensus of those who have studied the area in depth, and on reminiscences of those who were there. Many reminiscences are tinted by fondness, family legend, and wishful thinking. However, they too have their place, for human history is subject to these influences. Personal reminiscences, therefore, are captured within quotes here and duly noted and should be considered in their own important but unverifiable context.

Welcome to Bodie.

BODIE'S VIOLENT PAST

Bodie's past was violent even before the first Argonauts passed through. The very geology of the region is restless and full of upheaval. Bodie's rocks are of volcanic origin. The current theory (which geologists caution us is subject to change in the light of new discovery) is that an active magma chamber lay under the Bodie region, from which multiple volcanic vents formed. Those vents are known today as Bodie Bluff, Standard Hill, Queen Bee Hill, and a few others. Shortly after these volcanoes stopped spewing, about nine million years ago, mineralization laid down gold in their internal fissures. (For a more detailed description of these processes, please see appendix 1.)

Although the Bodie region has long been dry and inhospitable to humans, people have lived in and around the area for many years. Archaeologists estimate that humans arrived here about 5,500 years ago (Fletcher 1987, p. 3), and successive groups of people have kept the area inhabited, though sparsely, from then until now. The earliest residents did leave collections of obsidian flakes where they crafted tools and arrowheads and other items. More definite signs of their civilization may lie some distance under today's topsoil, perhaps even in downtown Bodie, under the vestiges of successive generations of humans.

THE KUZEDIKA

The Kuzedika (pronounced "koo-ZEH-dee-kah") were the people living in the region when W. S. Bodie discovered the first gold in what is now the Bodie Mining District. Although they were not the very first human inhabitants of the area, the Kuzedika people had been there for many generations by the time the first European-Americans arrived. Virtually nothing is known about the people who lived in the area before the Kuzedika arrived.

The name Kuzedika means "fly-larvae eaters" in the Kuzedikas' own dialect.[1] These people are better known for their custom of eating the kutsavi, or fly larvae, from Mono Lake than for their habitation of the Bodie area. The Yokut Indians to the west referred to the Kuzedika as the *monachi,* which is presumed to refer to the same custom, because the Yokut word for "fly" is *monai.* The Kuzedika also referred to themselves as *nümü,* which means, simply, "persons." Their language is part of the Shoshonean language family. They lived a hunter-gatherer existence.

When the gold seekers arrived in the Mono Basin, the Kuzedika population was very sparse, and, correspondingly, they did not have a history of major conflicts. No Kuzedika settlements existed in the Bodie area at the time. After Bodie was established and growing, more interaction between the European-Americans and the local Kuzedika people was inevitable. (For more information about the Kuzedika, see appendix 3.)

BODIE: THE MAN AND THE TOWN

Beginning in 1848 and 1849, with the explosive invasion of the '49er Argonauts (gold seekers), California's gold country in the western Sierra was infiltrated by men and, to a lesser extent, women from all over the world. Most were here after the precious "yellow" or "color," but some came to find their fortune servicing the miners and helping to build supporting communities. The mining action stayed on the western side of the Sierra for several years, until 1852, when Ahwahneechi Chief Tenaya fled Yosemite Valley with the U.S. Cavalry in pursuit.[2]

Chief Tenaya fled after being warned that his people were blamed for violence against local white settlers and that he would most likely be killed if caught. Lieutenant Tredwell Moore of the U.S. Army headed up a retaliatory expedition to search for the chief. In pursuit, Moore and his men crossed the Sierra over the Mono Trail through Mono Pass. On the eastern side of the Sierra his men discovered gold, gold-bearing quartz, and other minerals in what is today Lee Vining Canyon, just below Tioga Pass.

Upon returning to the Mariposa area (without having found Chief Tenaya), Lieutenant Moore displayed their find and published an account of the expedition.[3] Predictably, several prospectors made an immediate rush to the area, including one LeRoy Vining, whose name was shortly thereafter given to the canyon and, later, to the town that still stands near the canyon and Mono Lake.

Other prospectors followed, and soon the exploration of the eastern Sierra's gold potential was in full swing. Mining camps such as Dogtown and Monoville appeared, which today are only notations on maps, all dwellings and dwellers being long gone.

BODIE'S BONANZA

In July 1859, W. S. Bodie (or Body or Bodey; both the spelling of his name and the specifics of his identity are disputed) wandered into the general vicinity of Mono Lake with a few prospecting companions. He had been prospecting here and there for some time. In Sonora, California, he met a Terrence Brodigan, later a Bodieite who enjoyed his status as an acquaintance of the founding father. Bodie had most recently been in Monoville, just east of today's Conway Summit. He set off from there with at least one companion, looking for color, as all prospectors did.

The identities of Bodie's companions are somewhat disputed. Many accounts claim he was with Terrence Brodigan. Sources disagree: Apparently Brodigan's tales had many inconsistencies, and Brodigan himself was the main person who insisted he went with Bodie (Wedertz 1969). By all accounts, Bodie was with E. S. "Black" Taylor, who was described as being half Cherokee. In the summer of 1859, Bodie and Taylor, perhaps with others as well, came through what is now Bodie and found color in the form of a placer deposit on the eastern side of Silver Hill, the eastern central part of the Bodie Mining District, at what is today called Taylor's Gulch. They filed claims on their find.

In the late fall of 1859, Bodie and Taylor returned to their claim, built a makeshift cabin, and began to develop their find in earnest. Bodie and Taylor traveled to Monoville, then an active camp, for supplies. A winter storm set in with the area's characteristic fierceness, hithertofore unknown to them, and while trying to return to the cabin, Bodie perished in the snowstorm. Taylor barely made it to the cabin, warmed himself up for a bit, then set out to retrieve Bodie. He didn't find him then, nor later after the storm subsided. The winter of 1859-1860 turned out to be particularly treacherous. Not until the following spring did Taylor find Bodie's body. He buried Bodie where he lay—no tombstone, no death certificate. Bodie's death is in keeping with the miners' superstition that those who discover rich deposits die tragically.

Black Taylor stayed on in the area. Sometimes he was in Bodie, and sometimes in Aurora, about twenty miles to the east in Nevada, where he had an active interest in several claims. Taylor died in Benton in late 1861 or 1862 (*Bodie Standard* 29 October 1879) when hostile Paiutes attacked him in his cabin. It is said that his skull was passed around for some time as a memento of the era.

WILL THE REAL W. S. BODIE PLEASE STAND UP?

Mr. Bodie's exact identity is somewhat uncertain; accounts of his first name vary from William to Wakeman to Waterman, and spellings of his surname vary from Body to Bodey to Bodie. Mono County records show that the original mining claim in the area, the claim from which the mining district's boundaries were established, was the Body claim. Because this is the original claim in the area, located by the prospector himself, that spelling cannot be discounted. The mining district's original name was the "Bodey Mining District" (Mono County Records, Records of Bodey Mining District 1862), although the district was formed after W. S. Bodie's death.

During Bodie's boom time in 1879-1880, the founding father became a popular topic, especially when what were likely his remains were unearthed. Because there was no death certificate for Bodie (in fact, no records existed outside of personal reminiscences), an aura of mystery surrounded his identity. At that time, several people who claimed to have known him described what they knew about him. Their claims, not surprisingly, are somewhat contradictory. Some claimed he was Dutch ("a Mohawk Dutchman," said one account); some said he had an accent. They all seemed to agree he was around forty-five years old when last seen and had a wife and children. Most said he had two children, but a later account credits him with six. He was not a large man, they said, and was a respectable sort, although one man claimed he was the dirtiest fellow he'd ever seen (quite a claim for prospectors).

How many people have attempted to describe the personal appearance of the late Wm. Body, who never saw a hair of his head. . . . From all accounts, Body was rather under-size in nature, say 5 feet 7, and weighed about 170 lb.; had a dark, sandy complexion, and was of very slow and slouchy habits. (*Daily Bodie Standard* 27 October 1879)

Although several gentlemen named Bodie or Bodey have been suggested as the town's misfortuned founder, none have been proven. The most likely candidate appears to be William Smith Bodie, born in Scotland, immigrated to New York and then, perhaps, to the California goldfields (for further information, please see appendix 4).

Which Bodie is the real McCoy? The truth at this point belongs to the ages. The lingering mystery only typifies Bodie's history, which is one of uncertainty and exaggeration.

THE NEXT CONTESTANT, PLEASE

In the summer of 1996, Bodie State Historic Park (SHP) was visited by no fewer than three more Bodie/Body families, each of which suspected that W. S. was *their* ancestor. One family was from Scotland, another from Germany, and the third from the Poughkeepsie, New York, area. The Bodies of Scotland told of a family legend about a long-lost relative who disappeared from the family and later had a California gold town named after him (Shipley 1995). Perhaps one day, through one of these family trees, the real W. S. Bodie/Body/Bodey will finally be determined.

Visitors today can still visit the Miners' Union Hall, which now holds the park museum. This photo was taken in 1891. Courtesy the Bancroft Library, University of California at Berkeley

Bodie's Time Line

Bodie's enormous past can be outlined with a list of assorted times and events:

Pre-1859: Archaeological evidence suggests the presence of Kuzedika people and Sierra flora and fauna in the area. Many remnants of past inhabitants have been located in and around the Bodie townsite area.

1859: W. S. Bodie and companion(s) discover placer gold just east of the Bodie townsite, in Taylor's Gulch.

1860: W. S. Bodie dies in a snowstorm. Later that year, the "Bodey" Mining District is organized, with the "Body" claim as its center.

1861: Bunker Hill Mining claim established. Later renamed the Standard Mine, this was Bodie's biggest-paying mine.

1862: In November, the "Bodey" Mining District becomes the "Bodie" Mining District.

1863: The settlement of Big Meadows takes the name Bridgeport as it establishes itself as a logging, agricultural, and ranching area. Bodie Consolidated Mining Company, the district's first corporation, is formed.

1868: The Sonora Pass wagon road is opened (much of today's Highway 108 follows this same road over Sonora Pass).

1875: The Bunker Hill Mine cave-in exposes rich ore.

1877: The Standard Consolidated Mining Company (the backbone of Bodie's gold business) is formed. Other investors follow suit, forming other companies. On 22 December, the Bodie Miners' Union is organized.

1878: The Bodie Mine's unbelievable bonanza sparks a stampede. In June, Bodie Miners' Union building is built.

1880: San Francisco Stock and Exchange Board features over 50 Bodie mining stocks. Mono Mills lumber operation gets under way. Bodie's population reaches between 6,000 and 10,000, depending on which citizens are considered actual Bodie residents.

1881: Bodie Railway and Lumber Company railroad to Mono Mills is finished. The incidence of bonanzas and new claims falls off, high-living types of citizens move on, Bodie's decline begins.

1892: Bodie's first big fire: 64 buildings are destroyed. The world's first hydroelectric power transmission, from Green Creek to the Standard Mine in Bodie, begins: first plant of its kind in America.

1895: The cyanide process of gold recovery brings a small resurgence to Bodie.

1911: The Jordan Power Station is built, and electricity comes to Bodie homes until an avalanche crushes the station, killing all but one person.

1917: The Bodie and Benton Railway and Commercial Company railroad (formerly the Bodie Railway and Lumber Company) ceases operation.

1928: Treadwell-Yukon and Syndicate mining activity picks up with twenty-four-hour milling.

1932: The Great Fire destroys most of Bodie.

1942: U.S. government shuts down all gold mining because of World War II war effort. Town empties when the mines stop.

1962: Bodie State Historical Park (SHP) opens to the public.

1988: Galactic Mining begins feasibility project with open-pit gold mines and cyanide heap-leach processing as a long-term interest. Proposed project sparks debate over compatibility of active mining operation with SHP operations and objectives.

1994: Bodie Protection Act, merged with Desert Bill, passes in Congress.

1997: California State Parks, with financial assistance from the American Land Conservancy, purchases a 520-acre parcel of land adjacent to Bodie SHP for inclusion in the park. The purchase includes mineral rights as well as the actual property. The appropriations are unanimously supported by such contemporary powers as President Bill Clinton, California senators Barbara Boxer and Dianne Feinstein, Governor Pete Wilson, and the Mono County Board of Supervisors.

1998: In December 1998, the U.S. Bureau of Land Management (BLM) acquires the property and mineral rights to the portion of Bodie land held by the American Land Conservancy, making all of Bodie publicly owned.

2

I soon shall be in Frisco and there I'll look around,
And when I see the gold lumps there, I'll pick them off
the ground.
I'll scrape the mountains clean, my boys, I'll drain the
rivers dry.
A pocketful of rocks bring home, so brothers don't you
cry! —*"Oh California!" Gold Rush-era song*
(Arlen, Batt, Benson, and Kester 1995)

The Gold in Them Thar' Hills

Mining activity began relatively slowly in Bodie. The Bodie Mining District[1] was formed in the fall of 1860. The first claims and owners were duly recorded. W. S. Bodie's original claim was the center point of the Bodie Mining District, and the district spread out five miles in each direction from it. The district's rules laid down the procedures for establishing and maintaining claims. The rules stipulated different claim sizes for different types of land. Flat claims were the smallest, hill claims the next largest, and ravine and gulch claims the largest. Each claim holder had a required amount of work to perform on his claim, within a specified period of time. Required work usually entailed using arrastras and/or ore rockers, which were traditionally used with water. The Bodie Mining District records made allowances for the area's scarcity of water by noting that mining work was required "where there is sufficient water to work with . . . ore rockers." In November 1862, the district specified several more rules, with an emphasis on mining gold-bearing quartz, which had by then become the focus of the district.

Aurora, some miles down the road just across the Nevada border, dwarfed Bodie's mining-activity fame in the 1860s. At that time, Aurora was *the* riproaring mining town, and its Esmeralda Mining District attracted all kinds of people, mostly men. One of its more famous temporary inhabitants, Samuel Clemens, missed finding his fortune only by failing to complete his claim's work requirements within the allotted time. However, the rest of us profited greatly from his backup profession: writing under the name Mark Twain.

In July 1861, the Bunker Hill Mine was discovered in the Bodie Mining District. This later became the famous Standard Mine, the most steady and profitable of all the Bodie mines. Next year the San Antonio claim was discovered. It was later sold and renamed the Bechtel Mine.

This pattern of discovering and selling mines reflected the common distribution of mining activity of the time. Prospectors abounded, and they specialized in locating promising veins, but not all prospectors were miners. The business of mining and processing the ore was quite different from discovering the ore. More than a few prospectors preferred to stick to the discovery business, staking and then selling their claims. Prospecting also required a much smaller capital investment than did mining and milling.

THE DAWN OF BODIE'S CORPORATE AGE

It was difficult and expensive to mine and process gold in Bodie, and life there was never as easy as in more mellow climates. Miners and, more to the point, investors soon found teamwork more attractive than individual effort. In March 1863 the first corporate mining interest was formed at Bodie. The owners of several adjoining mines banded together to form the Bodie Bluff Consolidated Mining Company, with then–California governor Leland Stan-

In the times before mills, miners processed ore using arrastras like the one pictured here. Courtesy California State Parks, 2002

Home Sweet Home

"A miner's cabin . . . is a very primitive specimen of architecture," wrote J. Ross Browne in 1869. Roofs were anything from clapboard or shingles to sod or canvas. Browne saw roofs made from "flour sacks, cast off shirts, coats and pantaloons, all sewed together like a home-made quilt." Miners who built homes against the bluff carved a hole into the bank for a chimney, others improvised with what they could find, from stone and metal scraps to wood, bricks, and apparently even canvas. Said Browne, "Think of a canvas chimney! How an Insurance Agent would stare at it!" Browne describes in dusty detail the domestic tranquility of a typical Bodie miner home in 1869:

> Push open the rough board or slab door, and you have before you the social and domestic life of the honest miner. If the walls are a little rough, and somewhat smoked in the vacant spots, what matters it?— they are abundantly ornamented. The chinks are stopped with gold and silver croppings; pegs project from convenient crevices, from which hang old boots, shirts, flitches of bacon, bunches of onions, and sundry other articles of apparel and subsistence; rough clap-board shelves heaped with books, hardware, crockery and groceries abound at convenient intervals; a bedstead made of pine logs, with blue or red blankets over it, occupies one corner, or there may be a row of bunks ranged along a side wall, ship-fashion, to accommodate straggling or casual inmates. Frying pans, tin cups and a coffeepot hang over the fireplace, by way of pictures. But even pictures are no rarity in the miner's cabin. The battle-scenes in *Harper's Weekly* form the most artistic collections in the mining community. Entire walls are covered with them—whole houses are papered with them. You can sit on a three legged stool in any of these cabins and see the great rebellion or the impeachment of President Johnson acted over from beginning to end. (Browne 1869, p. 402)

ford at the helm as president and Judge F. K. Bechtel as secretary. The purpose of the organization was to finance the mining and milling of Bodie's gold by attracting outside capital. They offered 11,100 shares (which brought in $1.1 million) with handsome certificates that featured an artistic rendering of Bodie's own Isabella tunnel.[2] The company did not succeed for long.

The gold business was a feverish, jittery enterprise. Success was often

Prospectors. Courtesy the Bell Collection

fleeting, and investors were very fickle. Rumors and pronouncements by "experts" could influence the success or failure of a given company or an entire region. That same year, 1863, Professor William P. Blake, a respected mineralogist of the time, encouraged potential investors when he pronounced Bodie's prospects as "favorable for richness in gold."

In the fall, the Bodie Bluff Consolidated Mining Company combined with Bodie Consolidated #2, Isabella, Tioga, and the Rio Vista to form the Empire Company of New York. Officially incorporated in July 1864, the Empire controlled 38,000 feet of mining claims, mill sites, tunnel rights, buildings, and so on, and set its worth at $10 million, again trying to attract outside capitalists to the Bodie goings-on. The company bought a quartz mill in Aurora and moved it to Bodie. The mill had 12 stamps that pounded the hard rock into rubble from which the gold was extracted. Once the mill was installed in Bodie, the Empire Company enlarged its capacity from 12 stamps to 16 stamps. The Empire Company also moved toward developing support services for employees, which further encouraged the growth of the town of Bodie.

THE FALL OF THE EMPIRE

The Empire Company, despite its enormous investment in Bodie, never found a significant strike. It collapsed in July 1867, perhaps because it did not do enough exploratory work to ensure its continued acquisition of rich claims. This was a common mistake among mining companies, according to Professor Benjamin Silliman, a noted geologist of the time. The Empire's fall signaled the beginning of Bodie's first "death." In 1874, the Empire quartz mill, which had been purchased in Aurora for $45,000, was sold for $450 in delinquent back taxes.

The aforementioned Professor Silliman, from Yale, spoke about Bodie in Carson City, Nevada, noting the volcanic nature of the land and its geologically youthful history. "Probably no district in the history of mining possesses such a singularly interesting and perfect formation and one which is certain to induce capital, industry and intelligence to go to extreme lengths in the cause of development," he declared (Wasson 1878, p. 13).

BODIE TOWN'S HUMBLE BEGINNINGS

The early history of Bodie was strangely peaceful; according to tradition, no life was lost in personal encounters in the district previous to last fall, when a street duel occurred wherein the two principals were shot to death. Since then other sallies with sidearms have occurred with serious results, but as a general thing the town is not an unruly one. —JOSEPH WASSON (1878)

From 1859, with the arrival of Bodie's first two prospecting inhabitants, until the beginning of the Bodie boom in 1878, the town's population was sparse but ever increasing. As corporate mining operations grew, the town grew as well, attracting tradesmen as well as prospectors, miners, and assayers.

But in the mid-1860s, Bodie was more of a mining camp than a town. The population was almost entirely male, and the majority of them were miners. County records for 1866 show that five trader licenses and four liquor licenses were purchased. The next year, the number of trader licenses was the same, but the number of liquor licenses went up to 111.

The camp in the mid-1860s featured at most fifty residents, about twenty wood and adobe houses, and a boardinghouse. Most of these dwellings were up on the hill, closer to the mines than the Bodie townsite. The early residents were variations on a theme: miners and prospectors there to find a buck, although not usually a fast one. There were no stores, no restaurants, or other vestiges of an organized society. The settlers traveled to Monoville or Aurora for provisions, just as Bodie and Taylor had when they encountered the fatal blizzard in 1859. The first white woman in town was most likely Mrs. Robert Horner, née Marietta Butler (Wedertz 1969, p. 6). The Horners were among the first settler families in Bodie and had the first baby in town, born in 1865 "according to the 'oldest inhabitant'" (Wasson 1878, p. 24).

Real estate expectations were high. Much of the area was staked out, though not with any sort of systematic organization. The promise of gold was fueling a slow but steady development. In the mid-1860s, the Empire Company aimed to direct this development, and poured between $300,000 and $500,000 in 1860s dollars into the area, erecting a mill, bringing in machinery, building a boardinghouse, and paying a staff to run its interests there.

Farms sprang up near Mono Lake, Bridgeport, and in the Walker River Valley about this time. By 1868 or so, Bodie settlers could get fresh vegetables that were grown in the Mono Basin. This was a marked boost to locals' health; until local vegetables became available, "scurvy was a common disease" among the miners (Browne 1869).

A trip to Bodie in the early years was a rugged adventure. Writer J. Ross Browne prepared to visit Bodie in 1868: "I was assured that if an utter lack of accommodation for man or beast, and a reasonable chance of suffering from chilly nights and dusty roads, could be accounted among the luxuries of travel, I would not be likely to regret the trip." Browne wrote up his travels for *Harper's Monthly,* a major publication of the day. His vivid descriptions painted a picture of Bodie that remained the popular notion for quite some time.

Bodie's residents, the prospectors and miners, did not value living conditions as highly as the potential for finding gold. Their dwellings were fairly rustic, and some were literally slapped together. They lived in canvas tent cabins, roughly constructed wooden lean-tos, and some even made do with little cave-cabins, carved out of the hillside and fortified with wood and whatever else they could scavenge. Most of these "homes" were up on Bodie Bluff. The rugged weather of the region forced those men who planned to stay for any length of time to build more solidly. As more people arrived, more structured buildings appeared.

They kept warm with wood fires. These weren't huge blazing infernos to warm one to the core; they were just enough to get by, because wood was hard to find out there. Better save what you have, especially if you're planning on making it through the winter.

Rustic as the lifestyle may sound, many of these early miners had no great desire to live in a more "civilized" environment. In fact, many of them considered *this* environment truly civilized when compared to a more populated and urban scene.

Many items, including food containers such as this mustard tin, became building materials in Bodie.

THE BODIE MINING DISTRICT'S SECOND LIFE

The early mining activity had died down somewhat when, in May 1868, the *Esmeralda Union* of Aurora, Nevada,[3] reported that "Kearnahan, Olsen & Co. (of the Great Western mining claim) struck a small ledge a few days since, at Bodie, of extraordinary richness. They have already taken out about eight tons of rock which is said to literally glisten with the precious metals. . . . Altogether, the prospects of Bodie are very encouraging." Almost one month later the same paper stated, "Several miners in from Bodie report mining operations to be in a flourishing condition. There are but a few miners in this camp at present, and the greater portion of them have been employed taking out ore. . . . Some fine ledges have been opened lately and all are doing well." The following month it was reported that from 13 tons of ore they got $3,100 "and odd" dollars. Bodie wasn't dead yet.

WHEELING AND DEALING IN BODIE

In 1873, Leland Stanford, one of Bodie's original major investors, grew doubtful about the area's prospects and hired an "expert" to determine the feasibility of gold mining there. This "expert" told him there was no color (gold) more than 200 feet below the surface. Stanford visited the area, became discouraged, and said he'd take $500 for his entire holdings. He did, and moved his mill elsewhere. In 1879, Stanford would return to see Bodie as a boomtown and to rue the advice of his "expert."

After the Empire Company gave its last gasp in 1867, the company boardinghouse was looked after by William O'Hara. O'Hara was an African American man who had run the Empire's boardinghouse in Aurora before doing the same in Bodie. After the Empire pulled out, he ran it for himself. He extended credit to many down-on-their-luck miners, including two partners, Peter Essington[4] and W. Walker. In 1873, O'Hara accepted their claim to the Bunker Hill–Bullion Mine as payment for the room and board they owed him. This payment included, specifically,

1,200 feet of the "Bullion Claim" on Bunker Hill

1,000 feet of the "North Bunker Hill Claim" on Bunker Hill

1,000 feet of the "South Bunker Hill Claim" on Bunker Hill

600 feet of the "Essington Claim" on the eastern side of High Peak Hill

All water rights and privileges, "overshot waterwheel," three arrastras on Rough Creek near the Aurora and Big Meadows (Bridgeport) wagon road. (Mono County Records, Records of Bodey Mining District)

All this was noted as having sold for $1,824.72 in gold coin. At that time, no one knew what the potential of these claims was, and O'Hara did not consider himself very fortunate. Being at least sixty years old, he was not really able to mine the property himself and looked forward to selling it, which he found very difficult to do. Bodie was not the current hot spot. Just over a year later, in 1874, he sold it all back to Peter Essington with his new partner, Louis Lockberg, for $2,361.25 in gold coin.

That same year, Essington and Lockberg sank a 120-foot shaft with no luck. They were on the verge of giving it all up when a cave-in exposed a fabulous vein of ore. This attracted quite a lot of attention, and they set to work, processing the ore in arrastras. The next year, 1875, Essington and Lockberg sold their Bunker Hill Mine to the Cook brothers, Seth and Dan, for $67,500. The Cooks renamed it the Standard Mine, and it went on to become the most steady and profitable mine in the Bodie Mining District. When Bodie's fortunes came down to one major operation in the twentieth century, it was the Standard. The stamp mill that is toured by interested state park visitors today is the Standard Mill.

The year 1875 saw the birth of another corporate enterprise, the Syndicate Company, which would last well into the next century. In the summer of 1876, Ed Loose discovered the claim that became the Bodie Mine, the second most profitable mine in the district's history.

William O'Hara: An Extraordinary Ordinary Man

William O'Hara was an African American man who was born free in Ohio in the early 1800s. He began his career working on Mississippi River steamboats. While working the boats, he met and began working for William C. Ralston as his personal steward. When Ralston headed to Panama to assist in a fledgling transport company, O'Hara went with him. In the fall of 1851, Ralston, with O'Hara still in his employ, filled in as captain on a voyage to San Francisco. On 1 October 1851, when Ralston began his return trip to Panama, O'Hara did not join him. Instead, he joined the gold fever scene, going to Columbia, California. There, he and his wife ran the very successful Jenny Lind restaurant until 1859, and owned a nearby ranch as well.

The O'Haras were both well liked in Columbia. In August 1859, they held a grand party at their ranch celebrating the 25th anniversary of the British government's abolition of slavery in the West Indies islands. The party was well attended and was noted a huge success by the local newspapers. Slavery was far from abolished in the southern states of the United States at this time, and California had more than a few proslavery citizens: This party was a brave act. After Mrs. Charlotte O'Hara died in 1862, William headed for Aurora and, eventually, Bodie.

William O'Hara, when he died, was hailed as Bodie's "Foster Father" because after the Empire Company pulled out, he kept the boardinghouse going and helped several miners in the hard times just before the boom. As county records of the time show, he was an astute businessman. At one point, he owned the claim that became the bonanza Standard Mine, after he accepted it in payment for miners' debts. He later sold it back to miners for cash, and not too much later, the big strike was made. Without "Billy" O'Hara, Bodie might have shut down entirely (well, for a time, anyway). A successful businessman, respected and well liked by all wherever he lived, William O'Hara was quite different from the stereotypical image of an African American man of the mid-1800s.

THE RISE OF THE STANDARD COMPANY

The newly formed Standard Company started out milling its ore at the Syndicate Mill on the northern end of town. However, it plunged in quickly, getting its own mill up and running on Bodie Bluff by July 1877. The Standard Company mill was operated by a steam engine that was fueled by wood and water. The steam engine turned a 17,000-lb flywheel, which turned all the machinery in the building. The engine used 20 cords of wood per day, costing the company between $20,000 and $25,000 each year. This wood requirement alone would support more than a dozen woodcutters in town. The Standard Company was becoming a force to be reckoned with in Bodie.

THE BODIE EXCITEMENT

The "old-timers" in Nevada used the word "excitement" to describe the rush to a new mining camp. —GRANT H. SMITH (1925)

The *Great Register of Mono County* documents the veritable flood tide of new citizens that began in 1877. Although only the men (and probably not all of these) are listed by name and profession, some women were arriving as well. The men were predominantly miners, but arrivals also included saloon-keepers, carpenters, teamsters (wagon handlers), blacksmiths, barbers, and others—even the occasional musician.

With the sudden population explosion, there was a new demand for housing. A building boom commenced, and signs of a larger, organized society began to appear, the first of which was the establishment of communication services with the outside world. In 1877 Bodie became connected directly with the outside world with the establishment of a post office, telegraph office, and in-town newspaper.

The post office opened in April. The next month, the new telegraph connecting Bodie with Genoa, Nevada, sent its first dispatch: "Bodie sends greeting and proclaims to the mining world that her gold mines are the most wonderful yet discovered" (Wasson 1878, p. 23).

Fear and Loathing in Bodie

In the summer of 1877, Ed Loose and his two brothers, William and Warren, encountered a strong campaign of intimidation from the Standard Company regarding their claim (Loose [1979] 1989). The brothers, expecting trouble, holed up on their claim, well stocked with ammunition. When a party of unfamiliar men tried to sneak up on them, one of the brothers fired toward them. It was enough to make their point. Shortly thereafter, the Standard Company approached the brothers, and they peaceably worked out an honest deal that left everyone content.

The Mono County Bank of Bodie was organized in August 1877, with O. H. LaGrange as president and H. F. Hastings as cashier. The same month, a second bank, the Bank of Bodie, was organized, with William Irwin as president (he later served as a mining superintendent as well) and W. H. Pope as cashier.

In October, the first Bodie newspaper, the *Bodie Standard*, the self-proclaimed "pioneer journal of Mono County, published by Frank Kenyon," started as a weekly and went to triweekly with the eleventh issue.

The *Bodie Standard* furthered the town's sense of established community. Although it provided a "polite" outlet for journalistic gossip, it also offered a consistent forum for some legal matters. For example, in each edition one miner or another filed "notices of protest." The notice of protest was a way for a Miner Joe to inform Miner Pete he must stop working on the Poohbah claim because it was actually owned by Miner Joe. Miner Pete was further warned that if his work didn't stop, he would "be proceeded against according to law." This matter might otherwise be handled less formally or politely in a saloon or on the street.

The paper was far from dispassionate and frequently offered opinions and advice to its readers, as on 7 November 1877: "One reason for hard times is dishonesty. If all men would live within their means, not try to ape the style of their neighbors, and by faithful labor, at such wages as they can get, pay their honest debts, we would hear less of hard times." It is rare to find such a statement in a newspaper today!

Bodie was transforming from a mining camp into a town. By October 1877, Bodie was home to enough families and children to warrant its own school district. Bodie citizens successfully petitioned for a Bodie School District separate from the Bridgeport School District. Three months later, in January 1878, Annie Donnelly opened a school in her house. She taught fourteen children at first, but by the summer the school had outgrown its quarters. The Green Street schoolhouse was established in February 1879. The schoolhouse that park visitors view today was the Bon Ton Lodging House at that time. The original Green Street schoolhouse, located two blocks up the street from the Bon Ton, burned down, allegedly set on fire by a "juvenile delinquent."

Bodie was becoming home to more families and people who desired a more organized town life rather than a rough and wild mining-camp life. With this came a growing desire for some form of legal order. By 1877 Bodie had the long arm of the law in town, in the form of two justices of the peace, a sheriff, two deputy sheriffs, and a constable, J. F. Kirgan. Over the course of his job, Kirgan became something of a local celebrity. The majority of Bodie law officials were from one group of individuals. The person holding any specific job changed periodically, but the personnel almost always came from the same group. This group is rumored to have been not altogether much different from the scofflaws and criminals they arrested.

On 22 December 1877, the miners formed the Miners' Union and elected their first officers. The first president was Alex Nixon, who was later laid to rest in Bodie Cemetery after a barroom disagreement, perhaps a fitting end to the first Bodie Miners' Union president. The Miners' Union Hall, later the Bodie Museum, was built in June 1878.

In December 1877, the Standard Company set loose an excitement when it declared its fourth dividend of $1 per share, a bonanza for shareholders. Four months later, in April 1878, only about a year after incorporating, the company had produced a total of $1,002,160. The entire Bodie Mining District's gold production for 1877 was valued at more than $700,000.[5] The Standard dividend acted as a catalyst, turning the growing interest in Bodie into a fevered rush. Scores of people began arriving in Bodie with high hopes and often little else.

The grave of Alex Nixon, first president of the Bodie Miners' Union, can be found in Bodie's cemetery.

*A great many strangers come to Bodie daily to take a look at our mines. They all seem
to go away well satisfied that they have seen the gold king of the world.*
—Bodie Standard *(7 November 1877)*

Bodie's growth continued unabated through 1878, despite the fact that the
year began in the middle of a fierce winter that saw many people just trying
to survive in Bodie's unforgiving climate.

As the town grew, its priorities and values were reflected in which kinds
of establishments thrived and which never appeared at all. In February, a
visiting reporter from the *Reno Gazette* noted:

Bodie has a population of 1,500, about 600 of whom are out of employment, and of
which the latter number, not 250 would work if they could find work to do. There
are in town 17 saloons, 5 stores, 2 livery stables, 6 restaurants, 1 newspaper, 4 barbershops, 2 butcher shops, 1 fruit store, 4 lodging houses, 2 boot shops, 1 tin shop, 1
jewelry store, 1 saddle shop, 2 drug stores, 3 doctors, 4 lawyers, a post office, Express
Office, 15 houses of ill fame, 1 bakery, 2 blacksmith shops, 2 lumber yards, 2 stage
lines, the usual secret societies and a Miners' Union. . . . There are six good mines
and about 700 locations. . . . The average new arrivals per day is 10.

Note the absence of churches. Just the same, there were religious services
held in the Miners' Union building. The clergy of different faiths coordinated
their schedules of worship to accommodate everyone.

On 20 February 1878, Bodie had the first documented visit from its nem-
esis, fire. Sam Chung's King Street restaurant, bakery, and lodging house
burned after a fire started in a defective stovepipe in the kitchen roof.
There was no fire department to fight it and, moreover, no water to fight it
with. Volunteers halted the fire's progress by knocking down neighboring
buildings.

Three months later, in May 1878, Joseph Wasson's *Bodie and Esmeralda,* a
booklet describing the "important revival of mining interests in Bodie and
Esmeralda Districts," was published in San Francisco. His book painted an
interesting picture:

Excellent water from springs and wells pertaining, in town, is obtained sufficient as
yet for ordinary uses. . . . [A]n historic cabin existed on the site of Gilson & Barber's
store—said to have been Body [*sic*] and Taylor's, and also where the original miner's
meeting was held organizing the district. But these accounts must reconcile themselves, along with a variety of others encountered during the preparation of these

pages. . . . It is a wooden town, and in itself contains about 250 structures; some very presentable ones. About the mines proper there are about 100 additional habitations and shelters of one kind and another. (Wasson 1878, p. 22)

Wasson hailed Bodie as one of the two "most interesting districts in the annals of mining" and included details about how to reach Bodie, and prevailing fares. His booklet only further beckoned people to the high-desert town.

In June of 1878 the Bodie Mine had a big strike 250 feet underground. Miners were cutting the Bruce ledge, which was so rich ($2,000 a ton) they bagged the ore before hoisting it up, to save every speck of gold dust. There they found a vein, the Burgess, that was assayed at $3,000 per ton.[6] This stuff of dreams enabled the Bodie Company to issue a $3 dividend.

Each year the mining companies had to pay tax assessments that Mono County levied on them. In July, several mining companies challenged their most recent assessments. The supervisors of the Red Cloud, Syndicate, Spaulding, and Richer mines all said their assessments were too high and succeeded in getting them reduced. When the Mono County district attorney tried to increase the Standard and Bodie assessments (the Bodie's bonanza having just been discovered), the attorney for the two mines challenged the district attorney's right to make such decisions. The board dismissed the issue when the district attorney didn't show up to explain.

The Bodie Company's dividend bonanza added even more to the circulating lore of Bodie. There was money to be made, if not from gold, then from gold miners. Various swindling types began to be attracted to the area as well as honest folk, and several fraudulent ventures took place outside of the Bodie Bluff and Silver Hill areas. Professional gamblers became a regular presence at the gaming tables in town, something the local newspapers noted as a confirmation that Bodie had "made it" as a mining town. It should be noted, however, that the real high rollers who won big in Bodie were the San Francisco–based investors who financed the mining boom.

As the 1878 gold production passed the $2 million mark, the incoming stream of people quickened. By the end of the year, the average arrival rate of about ten people per day had built a respectably sized town. The Christmas Eve edition of the *Daily Bodie Standard* reported that the town population had reached "some 5,000 inhabitants." The *Mono-Alpine Chronicle* referred to Bodie as a "city," saying, "a place the size of Bodie should be entitled to be thus classified." And the rush showed no signs of slowing.

Bodie ... has arisen as if by magic on a broad plateau on the summit of the Sierras, and to day bears evidence of prosperity not equalled by any mining camp in California or Nevada. (*Mono-Alpine Chronicle* 7 December 1878)

THE BAD MAN OF BODIE IS BORN

> *I'm a tornado. I tear when I turn loose . . . I'm "bad"; I'm chief in this yer camp,*
> *an' I ken lick the man's says I ain't. I'm a ragin' lion o' the plains, an' every time I*
> *hit I kill. I've got an arm like a quartz stamp, an' I crush when I reach for a man. I*
> *weigh a ton, an' earthquakes ain't nowhar when I drop.*
> —*Bad Man of Bodie letting loose* (Argonaut *1 June 1878*)

On 1 June 1878, the *Argonaut* newspaper in San Francisco published the tale of the "Bad Man of Bodie." This is believed to have been the start of the legend that persists to this day. The article describes a meeting between a boisterous blowhard, Washoe Pete, "generally considered a 'bluffer,'" and a meek, mild "expert, a pale, small man," sent to Bodie on a real estate matter. After the "bad man" threatened the expert with strength and weaponry, the expert protested he had no weapons. The confrontation then came down to fisticuffs, at the bad man's insistence:

[The] expert "countered" on the "badman's" cheek, and then stretched him panting for breath on the floor with a "stinger," straight from the shoulder, inflicted upon the lower portion of the chest . . . and as the "bad man of Bodie" crawled away he was heard to mutter that he "didn't lay out to fall up against batterin' rams," no more'n he 'lowed he was game in front of a hull gymnasium.

Although the article itself exposed the dubious (and possibly mythical) Washoe Pete as a fool, the phrase "Bad Man of Bodie" (sometimes "Bad Man *from* Bodie") caught on and was bandied about for quite some time. Many men claimed to be the actual Bad Man of Bodie, and it may be just as well, because it seems Bodie had enough bad men that settling on just one would be impossible.

The part of town that once featured the saloons and brothels, bordered by Bonanza and King streets, is now just a field with one lonely structure left (noted in the park brochure only as "Chinese Residence"). But in 1878, this section of town was fast becoming dangerous. The combination of exhausted miners, alcohol, boomtown atmosphere, and guns made Bodie nights prone to violent outbursts. Rumors of daily murders spread quickly. The expression "Do we have a man for breakfast?" was claimed by some to have begun in

Bodie, referring to men killed in the night in one conflict or another. Bodie's "bad man" reputation grew about as quickly as the town and soon became a source of pride, as one can see from local press items:

That last little shooting scrape has given Bodie a very bad reputation—for marksmanship. The papers are poking fun at us all over the Coast. Listen to the *Oakland Radiator*, for instance: "They have some very poor marksmen over at Bodie, and some resident undertaker ought to start a shooting gallery there. About two o'clock last Monday morning two men emptied two six shooters at each other across the counter in a bar room with no other effect than the tapping of a barrel of ale, one of the men then retired to the street, where he obtained a fresh supply of ammunition and the firing was kept up until nearly daylight, putting three balls through the glass doors and shooting off a cigar in the mouth of a passing stranger until it was too short to smoke. The man who had his cigar shot off got mad and woke up the police, when the firing ceased." (*Daily Bodie Standard* 21 January 1879)

Clearly this was played for humor. As the population increased, the papers would report more heinous crimes.

1879: GOING STRONG

> *A prominent physician says that the town is distressingly healthy.*
> —Bodie Weekly News *(21 December 1879)*

In the boom times of Bodie, 1879 stands out as the year with the most new arrivals to the town. A Grass Valley newspaper (*Bodie Standard News* 25 December 1878, reprinted from the *Grass Valley Union*, quoted in Loose 1989) reported that "[t]he average arrivals are about thirty per day, and all departures intend to return. . . . There are 47 whisky saloons, 10 faro tables . . . 2 banking houses, 5 wholesale stores, an excellent daily paper, and all the accessories of civilization, and refinement will soon follow." Fueling the influx further, over the course of 1879 about $2.5 million in gold bullion was shipped out of Bodie.

Despite the encroachment of civilized living, Bodie residents held fast to elements of rough living. There were still no church buildings. There were no fire hydrants in case of fire. Water and wood remained scarce and extremely valuable commodities. Life was rough, as were the miners, but as of yet Bodie had seen neither a church nor a hanging, the two elements fabled by miners to destroy a mining camp. Times were good and getting better all the time.

The winter of 1878–1879 was exceptionally mild—for Bodie—and the customary huge snowdrifts were not there to slow the flow of aspiring citizens. Locals became concerned about so many people arriving with neither work nor shelter. They knew the winter weather was severely unpredictable. The 30 January *Daily Bodie Standard* recommended, to anyone who would heed, "We would advise all parties looking Bodiewards to defer their coming until later in the season, as we may have much bad weather yet."

WOOD: THE BODIE STAFF OF LIFE

In 1879, the boom brought more new businesses and buildings than seemed possible: "[N]o less than 20 or 30 new buildings started within the last week," the 6 August 1879 *Standard* announced. "Persons living in the upper end of town, on leaving their homes in the morning and returning at night, are greeted by the sight of some new building which has sprung into existence since their departure from home." Virtually all the buildings in town were constructed of wood.

The demand for wood and lumber was great. The mining companies were operating full steam ahead—literally. Almost all the machinery was powered by steam, fueled by wood-burning furnaces. The Standard Mill alone had a standing order for ninety cords of wood per day, up from the twenty cords per day it required in 1877. Part of this supply was stockpiled to prevent shortage in colder weather. Wood was also needed to "timber" the mines, which were unstable holes in the ground without the wooden supports, and slightly more stable holes with them. Throughout 1879 and much of 1880, there was a great deal of reinforcement work going on in many mines (the building of shafts and drifts, and the bringing in of machinery) for exploration and mining work.

The ever-increasing demand made for a brisk wood-and-lumber business in Bodie. This was a special entrepreneurial challenge. There is not much wood in the immediate vicinity, at least not enough to build much of a fire, let alone a town.

The vast quantities of lumber used in Bodie came from more than one source. In the 1870s, "huge loads" of lumber were brought to Bodie from N. B. Hunnewill's sawmill in Buckeye Canyon, near Bridgeport. There were other mills in the Bridgeport area that sold wood to Bodie as well, and still more mills as far away as Benton and Carson City.

Cordwood was used for fueling fires in the mines and mills and for home heating. Piñon trees, the source of the pine nuts so important to the Kuzedika way of life, were the favored source for cordwood. The piñon wood is full of

Firewood Theft

As winter dragged on in 1879, the problem of firewood theft was addressed repeatedly in the newspaper. Because the only source of fuel in Bodie was wood, most people laid in a sizable supply in anticipation of winter. A few, playing grasshopper to the diligent ant, preferred to steal wood from other people's woodpiles. This outraged citizens who found their supplies diminished unexpectedly.

When the local lawmen seemed unable to stop the problem, some citizens developed a dramatic, but nonconfrontational, solution to the problem: They filled pieces of wood with gunpowder and placed them seemingly at random in their woodpiles. It was a short matter of time before the entire town knew who was stealing wood. The thieves were left at least stoveless and, in some cases, homeless by the explosion. No loss of life was noted from this (*Bodie Weekly Standard* 4 December 1878, quoted in Johnson and Johnson 1967, p. 46).

pitch and burns very hot, which was useful for driving steam engines. There were several piñon-rich areas near Bodie, such as the Rough Creek region, and woodcutters harvested these areas. Different woodcutters favored different locales. Most European-American woodcutters considered the Rough Creek region too difficult to work profitably, "yet the Chinese cut thousands of trees on the steep, basaltic canyon walls. The wood was packed on mules to King Street, where merchants piled large cords" (Wedertz 1969, p. 36).

Some woodcutters hauled their wood to town in wagons pulled by teams of mules. Others loaded the wood onto their mules' backs and drove the long mule trains into town. The wood-laden "Bodie Bobolinks" were a familiar sight in Bodie.

BODIE TOWN LIFE IN 1879

In February 1879, the now-famous tale of a little girl's coming to Bodie was born:

"Good-by, God; we are going to Bodie in the morning," was the suggestive termination of a sweet little three year old's prayer the other evening at San Jose. . . . All right, pardner, but we have no particular use here for a god [*sic*] that confines himself to the

These wood-bearing mules were called "Bodie Bobolinks" in the town's heyday. Courtesy the Dolan-Voss Collection

limits of San Jose; and we don't wonder that even a little three year old was willing to say "good-bye" when she thought she had a chance to get out of that delectable place in order to come to Bodie. (*Bodie Standard* 18 February 1879, commenting on a story in the *Nevada Tribune*)[7]

On Monday, 17 February, the new schoolhouse rang its bell and opened its doors to about eighty students. The now-sizable gold town of Bodie was of interest to people everywhere, and papers all over the state carried stories of Bodie on a regular basis. In keeping with the journalistic mission of "*all* the facts," the *Benton Mono Weekly Messenger* reported on the front page that there were only two cats in Bodie on 19 April 1879.

The unusually dry winter's effects were keenly felt as the Fourth of July approached. The holiday had already been established as a big celebration in Bodie, a time when everyone celebrated with abandon, and with Roman candles as well. J. F. Kirgan, currently the deputy sheriff, cautioned all likely suppliers of fireworks to curtail their inventory for fear of fire—the town still lacked fire hydrants to fight fires.

Some months later, the owners of the Bodie Bank, ever concerned about protection from prospective robbers, proudly announced the arrival of their new fire- and burglar-proof vault. To give clients the utmost reassurance about the safety of their assets, the bank's cashier and two assistants slept in the vault at night.

By the end of 1879, Bodie was no longer just a mining camp. The more recently arrived families did not have the miners' tolerance (or even enjoyment) of a violent and lawless atmosphere. The frequent mishaps in Bodie were becoming a topic in the papers. Even though Bodie had a reputation for violence, and the level of gunplay was high, it was not viewed with a "devil-may-care" attitude by the entire population, as the following item illustrates:

The practice of firing off pistols in the streets at night seems to be on the increase. Every night from 2 to a dozen shots can be heard, and very often these shots are fired in such close proximity to dwellings that the inmates are seriously alarmed. We hardly know what to say about the men who do the firing. The practice is so utterly senseless that one cannot help wondering what makes men do it. If the parties were arrested and fined $100 or so they would think it a great hardship, but it is the only thing that will stop it. If the officers would use a little diligence and catch a few of these fellows who get funny with their pistols, and have them pay a good round fine or stay in jail for a couple of months, we would be annoyed a good deal less by shooting at nights. (*Weekly Standard-News* 21 December 1879)

On Christmas, the Noonday Mine staff fired up their new 30-stamp mill, to the delight of the entire town, despite the noise. Later that week, a snowstorm, which citizens of that time would always refer to as "the Great Storm," hit the town. Work stopped for several days because of the weather. Winter had returned to Bodie.

ALAS, POOR BILL BODIE

As Bodie grew to greater prominence and size, so did the legends and curiosity about her founder. When Joseph Wasson visited in October 1879, he, along with Judge J. Giles McClinton, was seized by a desire to find W. S. Bodie's remains. They rode out in search of the location E. S. Taylor had once described to McClinton, about a mile from town, and began digging. Amazingly, they found bones and clothing that matched the description very closely. Wasson and McClinton quickly returned to town, and the next day, with an honor party, carefully carted the remains back to town where they were put out for viewing. This was greatly enjoyed by various citizens who not only viewed but also personally inspected Bodie's bones. "The skull, which had been carefully cleaned and polished like a billiard ball, would be taken up and closely scrutinized as if it were a piece of quartz from some new discovery" (Johnson and Johnson 1967, p. 16).

Townspeople felt that poor Bill Bodie deserved the proper Christian burial that circumstances had denied him, and they fell to preparations with gusto. "The remains were placed in a handsome coffin about five feet long covered with black cloth, and decorated with neat ornaments. On the lid was a silver plate bearing the inscription: 'In Memoriam, W. S. Bodey [sic], 1879' " (Daily Free Press 3 November 1879).

W. S. Bodie's services were held on the afternoon of 2 November 1879. When the appointed hour arrived, the Masonic hall was filled with mourners, and the sidewalks were overflowing. Four pallbearers bore Bodie's coffin up Main Street, while the fire bell tolled solemnly. Flags were at half-mast. A large crowd waited at the cemetery, including Terrence Brodigan, "old Bodey's companion." After the coffin was lowered into the grave, the Rev. R. D. Ferguson delivered the eulogy.

He praised the relatively unknown Bodie for his bravery, stamina, and vision; he lamented Bodie's lonely death; he visualized Bodie prospecting still "on the other side of the great river," and finally, he laid him to rest: "Let him repose in peace on this lofty summit, . . . here, where he blazed the trail and marked the first footprints to our golden peaks. Let a fit and enduring monument be reared to his memory. Let its base be wrought from the chiseled granite of these mountains. Let a marble shaft rise high above with sculptured urn o'er-topping, with the simple name of BODEY [sic] there to kiss the first golden rays of the coming sun, and where his setting beams may linger in cloudless majesty and beauty, undisturbed forever" (Daily Free Press 3 November 1879).

But it was not to be. Two years after the rapturous speech and the creation of the fit and enduring monument, President Garfield was assassinated. In a fit of patriotic fervor, Bodie's monument was adapted for Garfield and erected in the cemetery. No one is sure where in the cemetery Bodie's remains are: They are, as before, unmarked.

3

O land of gold, you did me deceive,
And I intend you my bones to leave,
So farewell home, now my friends grow cold,
I'm a lousy miner, I'm a lousy miner,
In search of shining gold.
    —"The Lousy Miner," Gold Rush–era song
      (Arlen, Batt, Benson, and Kester 1995)

The Lousy Miners

Mining was, and often still is, a brutal way to make a living. Mining in Bodie was as rugged as it gets. Accidents happened, some particularly grotesque. Yet hundreds of men thought nothing of leaving a more stable and traditional lifestyle to pursue their fortune this way. They became miners, mill workers, hoisting engineers, assayers, and more; the mining business required many professions in its daily operations.

HOISTING ENGINEERS

This thing of being dropped down two hundred feet into the bowels of the earth in
wooden buckets, and hoisted out by blind horses attached to "whims" may be very
amusing to read about, but I have enjoyed pleasanter modes of locomotion.

    —J. ROSS BROWNE ([1865] 1981)

The hoisting engineers were important fellows; they got the miners in and out of the mines safely and brought the ore out with a minimum of spilling. Although it sounds straightforward, it is hazardous work. When things go wrong, they go very wrong.

BOB BELL

Bob Bell, born and raised in Bodie and on Bodie mining life, recalled, while talking with Ranger Jack Shipley in 1986, the intricacies of hoisting. They were standing among the ruins of the hoisting works, next to the SHP parking lot:

The Lousy Miners. Courtesy the Cain Collection, Bodie State Historic Park

My granddad [Lester E.] ran the . . . big old one on the Standard shed. . . . [T]hey used to call all those hoist men "hoisting engineers" and they got paid a little more money than the miners.

As they discussed the mechanics of the hoisting process, Mr. Bell recalled the precautions, both mechanical and human, that they took to ensure safety when hoisting people or ore in and out of the mine shafts.

Well, it had quite a bit of weight there, it probably had about three or four tons hanging in the hole . . . bring out one ton at a time. . . . They go down the cable hook here. . . . [If] something went wrong with the cable or the rope part, . . . the cage wouldn't be free to fall, that thing wouldn't move down [but] two inches and these safety dogs would grab and hold it. . . . [T]hose old things are pretty sensitive; if you'd feed down too fast, if you'd slack the rope quick, why, they'll grab on you . . . and when they grab, boy, they are grabbed!

. . . My old granddad, he said the big one on the Standard shaft would pick the cage up on the 1200 foot level and have it on top in one minute . . . a quarter mile a minute . . . feels fast, you know . . . it *is* fast. . . . That thing right there'd lift ten tons easy. My old granddad said the big one on the Standard shaft, they guaranteed that it would lift about sixty tons . . . [but they] never had any use for that weight.

A Visit to Bodie's Underground

J. Ross Browne visited the Bodie mines early on, in 1863. He chose to descend without the help of a hoisting engineer and, after, to ascend with their help. He described his venture into Bodie's mining netherworld in detail:

> The shaft was about four feet square, rough, black and dismal, with a small flickering light, apparently a thousand feet below, making the darkness visible. It was almost perpendicular; the ladders stood against the near side, perched on ledges or hanging together by means of chafed and ragged-looking ropes. . . . I seized the rungs of the ladder and took the irrevocable dive. Down I crept, rung after rung, ladder after ladder, in the black darkness, with the solid walls of rock pressing the air close around me. . . . When I had reached the depth of a thousand feet, as it seemed, but about a hundred forty as it was in reality . . . I was shaking like a man with ague. . . . I was now quite out of breath and had to cling around the ladder to avoid falling. . . . With a desperate effort I proceeded, step after step, clinging to the frail woodwork as the drowning man clings to a straw, gasping for breath; the cold sweat streaming down my face, and my jaws chattering audibly. The breaks in the ladders were getting fearfully common. Sometimes I found two rungs gone, sometimes six or seven; and then I had to slide down by the sides till my feet found a resting place on another rung or some casual ledge of rock. To Jansen, or the miners who worked down in the shaft every day, all this of course was mere pastime. They knew every break and resting-place. . . . By good fortune I at last reached the bottom of the shaft. . . . A bucket of ore, containing some five or six hundred pounds, was ready to be hoisted up. . . .
>
> "Stand from under, Sir!" said Jansen, dodging into a hole in the rocks; "a chunk of ore might fall out, or the bucket might give way."
>
> Stand from under? Where . . . was a man to stand in such a hole as this, not more than six or eight feet at the base. . . . The bucket of ore having gone up out of sight, I was now introduced to the ledge upon which the men were at work. It was about four feet thick, clearly defined, and apparently rich in the precious metals. In some specimens which I took out myself gold was visible to the naked eye. . . . This was at a depth of a hundred and 75 feet. (Browne [1865] 1981, p. 407)

A typical hoisting works.
Courtesy the Gray-Tracy
Collection

Mr. Bell recalled that there were times when things went wrong in the hoisting operations: "They do sometimes, if there's something sticking out in the shaft and the cage hits some piece of timber or something, and stops the cage. Then all your rope will slack, you can tell you've hit something . . . the safety dogs'll grab then . . . you know there's something wrong."

Ranger Shipley commented, "Bet if there's anybody in the cage, they're getting nervous . . ."

Bell replied, "They're getting more than nervous! They're supposed to stop you, you know. They've got signal bells. Anything goes wrong, why, if you're fast enough you can stop the hoist right now. . . . [There's] just a wire hanging down [to grab]. . . .

"You can stop him, you can ring him up, ring him down, ring him slow or you can ring him fast. . . . [T]hey got separate signals for a man. If there's anybody riding on the mine cage, . . . you always gave them a three signal first, that says 'man cargo'—somebody riding it. Don't get on and just give a cargo signal, because they'll go like hell. . . ."

Mr. Bell remembered being up at the hoisting works with a full ore car coming up in the cage, which required a second worker to help handle the car when it got up to the top of the shaft:

They didn't like having anybody riding [up] with a car. . . . [The] company always has a . . . "top man." He handles all the rock coming up . . . somehow it takes more than one, you know. . . . [T]hey just [hauled it] out to the end of the mine, . . . dumped it into those chutes, loaded it into a truck . . . regular truck. . . . They used to dump it into those big old bins back under the hill there, they could hold 200 tons. . . . There was a rail all the way from the mill down and all the way in 2000 feet into the chutes. They used mules to pull those cars. That tunnel runs uphill a little bit towards the shaft, so when you load it up on the inside [the higher end] it comes out on its own weight. . . . The big one [Bulwer] . . . was our main working tunnel . . . we used to use it all the time.

It's a three compartment shaft . . . a manway and a pipeway and [a cage run] . . . you can't have anything in the cage run, where your cages are going up and down . . . [pipeway is where the pumps run] . . . [the manway is] just a ladderway down, they always made 'em have ladders in and out so if you had to, you could climb in or climb out.

In the mine shaft with an ore cart ready to go. Courtesy the Cain Collection, Bodie State Historic Park

Ranger Shipley mused that it must have taken a long time to climb out, and Mr. Bell concurred:

An old fella'd never get out, he'd be down there yet! . . . [You'd] come up pretty slow, pretty slow. . . . There's a round cable, that was probably up there . . . the one they used last . . . it's been used, see the old rope tar? They'd always grease the cable with that old tar, tar and oil; it keeps the rust off . . . those old babies have made many a trip . . . always had a load on them. . . . They don't go too fast . . . you don't have to let it down with the engine; if you want to, you can let it down with the brake . . . and the weight of that big heavy cage and that cable, it'll go down there pretty fast . . . it gets down there a little bit faster than you'd like to ride it! . . . Make your hair stand on end! . . . I never did ride cages very much, but I rode 'em to know what it was like . . . when they start down and you start to go to 1000 feet deep, boy, about every 100 feet, why, you pass a station where there's lights and they just blink and blink and you wonder, after you'd pass about a half a dozen of them going down, and you're going down pretty fast, boy, you wonder if you're ever gonna hit the bottom. That feels like a long ways. It *is* a long way!

In J. Ross Browne's account, once he had descended a shaft via a ladder and visited the workers, he elected to return via the hoist, which was not without its adventure:

Having concluded my examination of the mine, I took the bucket as a medium of exit, being fully satisfied with the ladders. About half-way up the shaft the iron swing or handle to which the rope was attached caught in one of the ladders. The rope stretched. I felt it harden and grow thin in my hands. The bucket began to tip over. It was pitch dark all around. . . . I darted out my hands, seized the ladder, and jerking myself high out of the bucket, clambered up with the agility of an acrobat. Relieved of my weight, the iron catch swung loose, and up came the bucket banging and thundering after me with a velocity that was perfectly frightful. Never was there such a subterranean chase. . . . To stop a single moment would be certain destruction; for the bucket was large, heavy and massively bound with iron; and the space in the shaft was not sufficient to admit of its passing without crushing me flat against the ladder. . . .

. . . I felt my strength give way at every lift. . . . My only chance was to seize the rope above the bucket. . . . This I did . . . and was presently landed on the upper crust of the earth, all safe and sound, though somewhat dazzled by the light and rattled by my subterranean experiences. (Browne [1865] 1981, p. 16)

The assayer is the fortune-teller of the mining business. The miners bring the ore to the assayer, who tells them whether or not they've found a fortune. A properly done assay determines which metals exist in what quantity and with what purity in a given sample of rock. Assayers use their own specialized chemistry sets to test for specific metals. In Bodie, they were looking for gold and silver. A former Bodie assayer tells the story of the process firsthand.

Gunnar "Pete" Peterson was an assayer for the Roseklip operation in Bodie from 1939 until 1942. Peterson studied mineralogy and assaying through the University of California. He then apprenticed with an assayer, beginning as a helper and ending up as an assayer. Peterson started out in Bodie as a pipefitter, but eventually replaced the previous Roseklip assayer. When he visited the park in July of 1986 he explained the process of assaying to Ranger Jack Shipley:

We had an automatic sampler . . . feeding the ball mill, and then that material would go through what is called "Jones splitters"[1] . . . and the sampler itself would cut maybe 500 pounds or so out of the daily run. . . .

Then out of the 500 pounds that would go through this Jones splitter, several Jones splitters, you'd end up with only about 20 pounds. Then the 20 pounds I would press through what's called a "chipmunk crusher"[2] and crush it down fine, then run that through a Jones splitter three or four times until I'd end up with about 2 pounds. . . . I'd pulverize it in a pulverizer down to about 150 mesh,[3] and then take a little cloth and mix it up good. I'd take a little spatula and dip it into a little pan in what we call a pulp balance. (Peterson 1986)

The pulp was weighed to an assay ton. An assay ton is 29.166 grams.[4] . . . If you had a milligram of button [gold] in there, that would represent . . . maybe an ounce of gold to the ton of ore. . . . You'd weigh that (the assay ton) and put it in your crucible, mix it up with flux[5] and put it in your furnace—we had a great big muffler furnace[6] that held 20 gram crucibles—and you'd put it in there with a little salt on top to cover it so it wouldn't bubble over. [You would] just smelt it and . . . pour it into a mold, which is called a king—it's a little cone shaped mold—pour one after the other in there, let it cool. . . . When it cooled . . . the flux would turn to glass. . . . You'd knock it off and what you'd have in there'd be a lead button. You add what's called a "litharge," which is lead oxide, . . . to the pulp and mix it in.[7] That lead oxide would absorb the gold and silver out of the pulp and settle it to the bottom of the crucible. You'd check the crucibles in the muffler . . . when they quit bubbling, you knew they were done. You'd pour 'em into the molds and let them cool.

While you take these out and cool them, you'd put in a bunch of what's called "cupels." They're little dishes . . . made out of bone ash mostly, and you put them in the muffler and heat them while your pour is cooling. Then you take your buttons up and hammer the glass off . . . make little squares out of them. Then those little cupels would be hot and you'd put a button in each one: of course you'd know which one was which.

Those cupels would absorb the lead out of the buttons. Part of it would volatize into the atmosphere up through the smokestack and what was left would be a little button, called a doré, that was gold and silver combined. Then you'd weigh that and . . . put it in little porcelain cups with nitric acid and water, 5 to 1, 5 parts [distilled] water, one part acid, put it on a hotplate and boil it slightly—simmer it like. . . . The nitric acid would dissolve the silver on the gold and you'd have a little dark button there—it wouldn't look like gold—then you'd pour off the solution of nitric acid. You'd save it all because you're going to save the silver out of that eventually—we had big bottles to pour it in. . . .

That little button there, you put it in the muffler while it's still hot, after it's dry. Dry it off first on the hot plate. [Then] you put those little porcelain crucibles which have the button in the muffle . . . it gets almost red hot, cherry red, and the gold turns from dark brown to gold color. It drives off any excess acid or anything else that's on there. Then you have your gold button.

You weighed your doré and [now] you weigh your gold, then you know how much silver you have, the difference between the two.

On bullion bars we'd run between 65–70 pounds. It'd be from 3–5 percent gold[8] . . . it looked like silver bars. The bars were about 18 inches long, 1.5 inches thick and 3 inches wide.

We shipped the bars by mail! We'd put them in a mail sack, seal it up and the mailman would come and take it away! . . . No guards, no nothing. . . . We put one bar in each sack; you couldn't go over 70 pounds or the post office wouldn't take it.

MINERS

When W. S. Bodie and E. S. Taylor found the first Bodie gold, it was just the two of them and any equipment they could lug in with the help of a couple of pack animals. Gold-harvesting technology has changed a lot since W. S. made his discovery, and Bodie has seen most of those changes implemented in its mines and mills.

The placer gold Bodie and Taylor found was not what is commonly thought of as placer gold, which is usually found in river deposits. It was old broken-down outcroppings from Silver Hill. In August 1859, Bodie discovered gold-bearing quartz veins on what became the Montauk claim, later renamed the Goodshaw. And they were off!

As the Bodie Mining District was developed, more and more shafts were dug throughout the district. A map of the existing shafts and workings looks like a confused honeycomb of crisscrossing tunnels and holes. Mining in Bodie in those days meant going down into the earth, blasting and digging the rock out of the mountain, hoisting it out to the surface, and extracting the gold and silver.

The standard warning cry when they were about to blast was "Fire in the hole!" This was everyone's cue to take cover from falling debris.

The mines were dark, and they could be pretty warm as well. Marian Hitchens Bryant recalled her father's tales of life in the mines: "One of his cousins once set a clutch of turkey eggs in a niche in the mine and they hatched . . . an idea of how hot the mine was" (date unknown, letter to Bodie SHP).

The general methodology for retrieving ore from the mountain remained unchanged for a long time. The last full-time operations in Bodie, in the

Fire in the Hole

J. Ross Browne, on his 1863 visit, was impressed by his descent into the mine. Having arrived at the first level of the operations, he was coaxed to venture lower into the shaft, which they covered with planks to protect the workers below from things falling from above:

> I had barely descended a few steps when the massive planks and rafters were thrown across overhead, and thus all exit to the outer world was cut off. There was an oppressive sensation in being so completely isolated—barred out, as it were, from the surface of the earth. Yet how many there are who spend half their lives in such places for a pittance of wages which they squander in dissipation! Surely it is worth four dollars a day to work in these dismal holes. . . . I scrambled down the rickety ladders til the last rung seemed to have disappeared. I probed about with a spare leg for a landing-place, but could touch neither top, bottom, nor sides. . . .
>
> "Come on, Sir," cried the voice of Jansen far down below. "They're agoing to blast!"
>
> Pleasant, if not picturesque, to be hanging by two hands and one leg to a ladder, squirming about in search of a foothold, while somebody below was setting fire to a fuse. . . .
>
> "Mr. Jansen," said I in a voice of unnatural calmness, . . . "How far do you expect me to drop?"
>
> "Oh, don't you let go, Sir! Just hang on to that rope at the bottom of the ladder, and let yourself down."
>
> . . . The ladder, it seemed, had been broken by a blast of rocks; and now there was to be another blast. . . . The blast went off with a dead reverberation, causing a concussion in the air that affected one like a shock of galvanism; and then there was a diabolical smell of brimstone. Jansen was charmed at the result. A mass of the ledge was burst clean open. He grasped up the blackened fragments of quartz, licked them with his tongue, held them to the candle, and constantly exclaimed: "There! Sir, there! Isn't it beautiful? Did you ever see anything like it? —pure gold almost—here it is! don't you see it?" (Browne [1865] 1981, p. 411)

A trio of miners at work. Courtesy the Bell Collection

1930s, were still digging the ore out and hoisting it up and out with mechanical hoists and rolling it out in ore carts pulled by men or mules. They were also going through old tailings piles with the cyanide process, but that story comes later.

Miner Marion Raab once told his relative Bob Sprague that when he was mining in Bodie, in the late 1920s and early 1930s, miners were required to undergo a physical examination before commencing work. The doctor who examined Raab asked him if he had a fondness for a lot of candy or a lot of liquor. The doctor commented that in his experience if a man did not enjoy sugar or alcohol to excess, he would not be able to withstand the rigors of the mining life, but he had no idea why it was so.

Everyone in Bodie was aware of the hazards of mining work. When tragedy struck a miner, the town's heart went out to him and, especially in later years, to any family he had there. When there was trouble in a mine, you could expect the whole town to turn out to help. As went the mines, so went the town. Correspondingly, it was expected that miners remain mindful of the safety of their fellow workers. Some demonstrated unusual heroics under stress and were lauded forever after. One of these was Elrod Ryan, who, in February 1879, slipped on an icy bucket that was used to transport men down the shaft. He fell 450 feet to his instant death. Remarkably, he called out as he fell, "Look out below, I'm coming," which everyone heard, enabling them to get out of the way and not be killed along with him. His consideration in such a dire situation was hailed in the local press.

GOING FOR THE GOLD

Once the rock is brought out of the mine, the job is to get the gold out of the rock. All methods for this, from ancient times to today, involve breaking the rock up into smaller pieces and then extracting the gold somehow. There are a few different ways to do this, and most of them have been used in Bodie.

In the early days, the processing equipment was fairly rudimentary. They used tools such as arrastras that could be operated by individual miners, or two- or three-man teams working on their own. They were powered by men or by one or two animals. After mining corporations began to form, the companies built stamp mills, which were powered at first by steam and later by electricity. Although mill technology changed and improved some over the years, the basic approach remained the same and, indeed, is used even today in some mining operations.

Both arrastras and stamp mills usually used mercury to extract the gold. Later and into modern times, a different process, based on cyanide, became more common. Each process is most effective at a particular richness of ore. For example, ore that does not work well with the mercury process because it is not rich enough may work more effectively with the cyanide process. Ore that is more concentrated than the process favors won't work as well either; you won't get as high a percentage of the gold out as with another process. As with all chemical reactions, what goes on and your end result depend on the balance of the chemicals you use.

Bob Bell recalled working on the amalgamation tables in the stamp mill in the 1930s. When they had too much of a good thing—rich gold ore—the process didn't work as well:

If you were getting a lot, you'd do what you call "rubbing up" every hour. . . . That quicksilver coating that's on there, starts to get kind of dirty and then it quits catching . . . [it] gets oxidized on top and it doesn't catch the gold as well as if you keep it stirred up, you know . . . so you take an old broom and just rub it around and break that scale up and make it all nice and shiny again. It starts to getting too much on there, that vibration keeps working it off the lower end, it keeps crawling down the plate, so if you get too much, why you just take some off, keep pushing up towards the top. . . . A lot of stuff stayed in the stamp battery . . . those old dies in there, they

The slurry ran over these amalgamation tables. The metal plates were painted with mercury, which caused the gold in the slurry to stick to them.

Shaking tables in the Standard Mill. The stamps at the top of the tables pounded the ore twenty-four hours a day.

just set 'em in there loose, and that gold gets in that loose iron and works down underneath it and it stays in there, too . . . you just had to get down in there and . . . take all those big old cast iron, big crushing dies out of there so you'd get at it, just scrape out everything. . . . The big company, they don't clean it up too good, cause they're gonna use it again. Ain't gonna go anyplace, it'll stay in there. Somebody'd had 100 tons of stuff, 50 tons of stuff, they would really clean it up . . . get every fraction of an ounce they could get out of it. (Bell 1986)

GEORGE DALY, PART I: THE MECHANICS' UNION STRIKE

Trouble in the mines was equally noteworthy in Bodie, and 1879 had two particularly important incidents involving disputes in the mines, both of which involved one George Daly.

On 13 February 1879, the hoisting engineers of the Mechanics' Union went on strike, demanding an eight-hour shift and a wage of $5 a day. This was a radical proposal: standard working shifts were twelve hours, and Bod-

J. S. Cain and friend display the Miners' Union flag. Courtesy the Bell Collection

ie's miners earned $4 a day, which was considered a high wage. The town was riveted for days as work ceased in all but a couple of the mines.

According to the *Bodie Standard* reports (13–15 February 1879), most of the mining superintendents immediately closed down works in response to the strike. The notable exception was George Daly, the superintendent of the Mono Mine, who told a reporter he "had placed himself in a state of defense." The town streets filled with men who would ordinarily be at work, the saloons stayed open, and the mining superintendents met hastily. The flag was left flying at the Mono Mine, a gesture that sparked much ill will toward Superintendent Daly.

On the morning of 14 February, several men approached Daly in town and asked him to attend the Mechanics' Union meeting. When he declined, they attempted to grab him. Daly produced his revolver, declaring he would "kill the first man who attempted to lay his hands on me." He then retreated into the Gillson and Barber store, where he hid out for several hours.

Eventually the Miners' Union interceded and put a stop to the strike without granting the hoisting engineers their demands. Because the miners greatly outnumbered the engineers, the engineers had little choice but to end the strike. The night after the strike was settled, several men appeared at the Mono Mine looking for Daly. "He was not present, a fact which he does not very much regret," the newspaper reported. Daly went back to work, but he had sown the seeds of resentment against himself and would have quite a crop to harvest before the end of the year.

GEORGE DALY, PART II: THE JUPITER/OWYHEE DISPUTE

Six months later, in August 1879, George Daly once more became the focal point of animosity in what is referred to as "the Jupiter/Owyhee dispute." (The following narrative relies on Loose [1979] 1989, McGrath 1984, and Wedertz 1969.)

The Jupiter and the Owyhee mines were located right next to each other. On 8 August, the Owyhee miners began sinking a shaft on land that the Jupiter crew claimed was theirs. On 11 August, Daly, now the superintendent of the Jupiter, told the Owyhee miners to stop their work. The miners replied that their claim was staked properly; he could check with the surveyor. Accounts vary, but all agree that both groups threatened to use guns to prove their point.

Daly returned a couple of days later with C. L. Anderson, the local deputy U.S. mineral surveyor. Anderson found that Daly was correct; the disputed land did indeed belong to the Jupiter. He also noted that a marker he had placed at an earlier time was now missing. Daly told the Owyhee men he would reinstate the marker and guard it with armed men. Daly then spoke with John Goff, one of the Owyhee owners and a popular member of the Miners' Union. Goff said Daly could buy the property for $4,500 or take them to court. Daly replied, perhaps unwisely, "Ammunition and guns are cheaper than law."

Daly returned on the evening of 22 August with twenty-five armed men, took over the Owyhee area, filled the shaft with dirt, and built a barricade. The Owyhee owners tried unsuccessfully to take back the area by force, operating from a miner's cabin about 200 feet south of the shaft. A small gun battle ensued, which lasted about an hour. Early the next morning, 23 August, the Jupiter men broke into three groups and attacked the miner's cabin that housed the Owyhee men, catching them off guard. John Goff was killed. The Jupiter men marched the surviving Owyhee men into town and turned them over to the deputy constable, who refused to arrest them.

People in town heard the gun battle, and word quickly spread about Goff's death. The townspeople grew angry. They blamed Daly and began to threaten him. People's anger focused on George Daly because he had instigated the Owyhee shaft takeover, although lingering resentment about his behavior during the earlier strike probably did not help him, either.

That afternoon, the Miners' Union met and decided to take over the

Miners resting after work. They worked hard: twelve-hour shifts, six days a week. Courtesy the Bell Collection

area in question. More than 500 armed men marched up the hill, joined by another 500 or so nonunion men. Before they reached the shaft house, they fired warning shots in the air. The Jupiter men fled. The foreman of the Mono Mine approached the men and was beaten. The mob burned the barricade and returned to town.

Accounts vary as to whether Daly simply left town (McGrath 1984) or was arrested and taken to Bridgeport with "a strong escort of deputies," where he posted bail and left for Carson City (Loose [1979] 1989).

The coroner's jury charged the six Jupiter men with the killing of John Goff. Daly and the other Jupiter men began their court trial on 10 September 1879, in Bridgeport. Patrick Reddy, famous as a defendant's dream, was their lawyer.[9] Most of the testimony supported Daly's allegations. All were acquitted of criminal charges. The grand jury also examined the case, and they, too, found Daly and his men innocent.

Daly returned to Bodie to manage the Jupiter mine once again. However, a month later the Miners' Union resolved to banish Daly along with the others responsible for the skirmish, ordering them to leave town within twelve hours. The resolution was posted in town. Daly answered by publishing a note in the *Standard*. His note emphasized two points: A court of law had found them innocent, and another court had determined that the original area of dispute did indeed belong to the Jupiter. Daly and his men then barricaded themselves in the Jupiter and prepared for war.

At this point, the locals became very concerned that a bloodbath would ensue. A citizens' committee was formed and met with Miners' Union officials and Daly. After a lengthy discussion, Daly agreed not to fight, and the union relaxed its terms, allowing the men forty-eight hours to leave and twenty-four hours in town once a month with no trouble to them.

Although Daly left, he and others publicly declared that the Miners' Union, not the law, dictated how things were run in Bodie. No one had brought the sheriff in during the Jupiter-Owyhee dispute, leaving the Miners' Union to act as judge and jury in the whole incident until a man was killed. After the topic was considered in the papers and among the citizens, public opinion in Bodie shifted to Daly's side, and he became a bit of a folk hero, although he never returned permanently to Bodie.[10]

4

*[C]onditions were . . . not what they have been
described for the sake of action or color. An excellent
description of the whole picture is contained in the
following: An Easterner approached an old resident of
Dodge City and said, "You don't seem to have as many
shootings and killings as you once did?"*

*Old Timer: "No, we don't, and what's more,
we never did."*

—GUY GIFFEN (1940)

The Golden Time

The golden time of Bodie was not terribly long-lived: 1879 through 1880. At that time, Bodie, California, was the current golden town of the Golden State. The gold business was flourishing, the gold miners were gleefully toiling, the merchants were gleefully taking the hard-earned gold pieces, and there was no end in sight to the good times. In 1880 Bodie's Standard Mine was called the "largest mine in the world." Plenty of rich veins had been found, and everyone expected more to be discovered. By 1880 (and for some time preceding) there were more than thirty Bodie mining companies on the San Francisco Stock Exchange.[1] During the course of the year, over $3 million in gold bullion was shipped out of town. And that figure does not include the gold that left unofficially.

Despite the natural hardships of the Bodie region, life anywhere in America in 1880 didn't get much better than what the successful citizens of Bodie enjoyed. It was a town with astounding amenities in an unlikely place and must have been an interesting place to live. To get an idea of the reality of boomtown Bodie, we can examine the town the same way we would explore a possible vacation destination: what it looked, sounded, smelled, and felt like; the people who lived there; the activities that went on; the attitudes and expectations the people had. There may have been gunfights in the streets and stagecoach holdups down the road, but what was it like to do the shopping, earn a living, and gossip in a town as talked about as Bodie? Did the Bad Man of Bodie do laundry as well as drink and shoot?

You could arrive in Bodie on one of several daily stagecoaches, on one of three different lines that serviced Bodie via Bridgeport or Aurora. These coaches traveled over privately owned toll roads, whose toll collectors lived in tollhouses along the road. The ruins of one of these, presumed to be toll collector Hank Blanchard's house, are still visible on the road to Aurora, looming like an old Roman ruin in the desolated countryside.

When you did arrive in Bodie, it was down one of the widest main streets known of any American town. And even though it was wide, it was crowded. There were twenty-mule-team freight wagons, ore wagons hauling rock to the mills, wood wagons and mules laden with wood bundles, hay wagons and lumber wagons, stagecoaches, people on horseback, and people on foot. Businesses lined the almost mile-long Main Street, and various supplies were stacked in front of the shops. As a small boy, Perry Buckley (a former resident) was awed by the sight: "He spoke of the coffins lined up along in front of the stores," his daughter, Mrs. Ken Warren, told Bodie SHP staff.

For accommodations, you had your choice of at least a dozen hotels, ranging from the cramped boardinghouses favored by impoverished miners to elegant hotels on a par with the finest San Francisco had to offer. Stepping out to dine, you had several choices to consider: There were the saloons with their chop stands featuring nineteenth-century fast food, or you could dine in a restaurant. These ranged from simple, good home-cooking establish-

Bodie's Main Street was crowded with wagons and citizens. Courtesy the Gray-Tracy Collection

ments to more elegant hotel dining rooms and restaurants with such unlikely gourmet offerings as fresh oysters. To this day, the shells from this luxury dot the outlying areas of the Bodie landscape.

The meals served up in Bodie tended to be hearty, starting with breakfasts that included steak, eggs, and hotcakes, with canned fruit and layer cakes laid out on the table for diners to help themselves. All this was made on-site and cost about a dollar per customer. You could of course wash it down with a variety of drink, but Bodieites were (and still are) proud of their fine water, which came from Rough Creek Springs, about four miles west of the town. "Bodie water is the best water in the world," says each and every one of the former Bodieites interviewed for this book.

The smells of Bodie were varied and plentiful: tempting smells of meals being prepared, from bakeries and chop stands to the exotic spices of China-town; freshly cut wood from sawmills; hot metal and fire from blacksmiths' shops; the ubiquitous livery stable aroma of hay mixed with livestock. On some days, with the proper wind direction, the saline essence of Mono Lake wafted in, giving Bodie the odd sensation of being a seaside town. And always in the background, the sagebrush with its reassuring herbal scent leaping forth after the rain, to remind all of its abiding presence.

Annie Miller's Occidental Hotel served up tasty meals each day. Courtesy the Cain Collection, Bodie State Historic Park

Bodie remained lacking in some forms of refinement, which perpetuated its air of rugged lifestyle alongside fancy city life. For example, there was no town drainage system, just dirt ditches here and there in an effort to afford some flow of liquids. And although the papers called repeatedly for the construction of sewers, none were built during the very populous years. Bodie's climate and temperature kept the issue from overwhelming the town, but there were undoubtedly health hazards to consider. When the snowmelt commenced, the streets were a gauntlet course of mud holes to be navigated around, much as they are today. But in those days there were also thousands of citizens and plenty of horses, buggies, and wagons traveling the streets each day.

The same Bodie streets that were swimming holes in spring were ankle deep in dust in summer, dusty enough to be watered down each morning. The 15 July 1881 *Bodie Standard* notes, "The street was not sprinkled this morning. What's the matter, Captain Gregg?"[2] Keeping the invasive dust out of the houses was a constant struggle on warm, dry Bodie days (and still is, according to SHP staff).

Once you had settled in, you would become familiar with the sights and sounds of this lively and noisy town. The stamp mills ran twenty-four hours a day, so there was always a background rumbling and, some say, trembling in the ground from the powerful machinery pounding up the rock to get at the elusive "elephant" therein (see appendix 1 for a discussion on the term "elephant"). The constant coming and going of wagons, animals, and people in the streets contributed to the background symphony—a far cry from the enveloping quiet of the townsite today.

You would see miners and mill workers trudging up to the mines and mills at the start of their twelve-hour shifts. In 1880, each of them earned a startlingly high wage of $4 a day, and they felt fortunate to earn such a high salary. Some of the people in town would, of course, have tales of instant riches to tell you about. Just before the boom, the mining companies were barely getting by and paying some people with stock certificates instead of cash. When the big bonanza happened, several lucky Bodieites cashed in their stock certificates for a good return. A few miners became wealthy overnight, and a few laundry workers earned ten times their usual fee because they'd been paid in stock. These tales contributed to the lucky atmosphere of the town, even though by 1880 it had been a while since any such lightning bolt of dramatic good fortune had struck.

You would find about 450 businesses in Bodie, including grocery and fruit stores, barbershops, hardware stores, blacksmith shops, millinery shops, and so on. Bodie also had the standard collection of doctors and lawyers as well as dentists, repairmen, photographers, and the like. Stuart and Sadie Cain (1977) said Bodie had three slaughterhouses. They remembered the high piles of dried bones that were hauled to Hawthorne, then shipped to San Francisco for use in sugar refining.

You would note that many women were in business in Bodie, and not just the infamous prostitutes. There were milliners and seamstresses, and women who were wives and mothers and cooked meals for miners and others in their legitimate boardinghouses. Others were house cleaners or clerks in shops. Some, generally considered to be fortunate, were housewives who ran their homes quite seriously, trying to keep a clean house in a dusty desert town, and raise children properly in the midst of many less-than-pristine examples of human character.

Although Bodie's popular image was not that of a family town, the two-story schoolhouse was filled with young scholars. There were 366 children in the three-year-old Bodie School District, although fewer than one-third of them attended school. Some spent their days working at home, or, the older ones, in stores or mines. Still others merely hung around town, absorbing a different kind of schooling.

When you were ready to relax and enjoy some form of amusement, you would find you had several options in Bodie, not all involving dissipation. You could watch (or participate in) any of several athletic competitions that were often held. These ranged from wrestling matches to walking contests to horse racing to baseball, all of which were very popular. Various lectures, performances, and dramatic presentations were given in town, often in the Miners' Union Hall. These included musicians and/or singers, variety shows, and serious drama. Some featured local talent; others were given by traveling performers. Periodically, one organization or another put on a ball, gener-

With the arrival of women and families, Bodie showed signs of "civilization."

Bodie had a baseball team for years. Ed Gray, Fern Gray Tracy's father, is the first on the left in the front row of this photo. Courtesy the Gray-Tracy Collection

ally in the Miners' Union building. The ladies in town particularly enjoyed these events.

Katie Conway Adair (1987) remembered her parents attending the dances and their tales of good times to be had there:

There wasn't anything allowed at the dances at all, but you could go next door if you wanted some wild life: to the saloons you know, 'cause they'd be gambling and that sort of thing. . . . I suppose people would come over when they'd been over next door, you know . . . if they were intoxicated they couldn't come in. . . . They were real cautious about their floor, you couldn't bring food in and spill it around or anything like that . . . they kept that place so beautifully. . . . The dances were beautiful . . . people kept time, the ladies and gentlemen were all so beautifully dressed. My dad had a white vest and he had diamond studs in it. . . . People who went to the dances were really dressed. I guess they wouldn't let them in if they weren't. I remember the shoes had to be "just so."

Reading was another popular pastime in Bodie. But it would be several years until Mrs. Annie Miller would open the first "circulating library" in her variety store next to her Occidental Hotel.

To keep up on local news, you could consult the "Personal" column in the *Daily Bodie Standard* to learn who was sick, who was in town visiting from other places, who had lost loved ones, and which famous people were coming to Bodie. The *Daily Free Press* offered a thinly disguised gossip article, "It Isn't True . . . ," that made oblique references to the goings-on of various townspeople as well.

The papers also kept you informed about those citizens who needed or received help. These stories describe a town that pulled together when solid citizens had a catastrophe. Benefit functions were held to help out those who had suffered calamities and found themselves in desperate straits. In the *Daily Free Press*, on 24 April 1880, such an event was advertised:

Miner's Union Hall should be crowded this evening. The appeal is made in behalf of a gentleman who is worth of the sympathy and aid of this community in these, his darkest hours. Although Mr. Ferguson feels his position keenly, and pride keeps back everything that appears as though he were in need yet the fact is apparent that he must leave for a warmer climate if life is to be preserved any great length of time. The performance will be a grand one, so let all turn out.

Bodieites enjoyed their leisure time in many ways, including carriage rides. Courtesy the Bell Collection

Hot Times in the Old Town

Bodie rarely overlooked an opportunity for merriment. Thanksgiving, Christmas, and New Year's always meant parties and balls and programs for people of all ages. Few holidays passed without the townsfolk finding a way to celebrate them. Such holidays as Cinco de Mayo, Bastille Day, the Chinese New Year, and the Paiute Indian Fandango did not pass unnoticed, either. The newspapers of the day describe various festivities taking place, and by all accounts townspeople of all backgrounds seemed to enjoy them.

Fourth of July, 1880

Of all holidays in Bodie, the Fourth of July was the grandest. People prepared for weeks for the celebrations, which sometimes spanned more than one day. In 1880, Bodie was an excited, prosperous town, and this was the citizens' chance to throw the biggest party the town would ever see. The newspapers describe the celebration in detail. Because 4 July fell on a Sunday that year, the citizenry waited until the fifth to celebrate. And then they did it in style.

At midnight, as it became the Fourth, thirty steam whistles sounded out. At 4:30 A.M., a brass band serenaded the town from the Belvidere Iron Works with patriotic songs ("Hail, Columbia," "Star-Spangled Banner," "America," and "Red, White, and Blue"). At sunrise the Standard's cannon was fired. The rest of the Fourth itself was quiet, as was traditional for Sundays.

On the fifth, after the Standard Company's artillery sounded off early in the morning, most of the town turned out eager for the celebration. Main Street had been decorated with flags, festoons of bunting, and evergreen trees lashed to posts. They transformed the dusty desert of Bodie's Main Street into a tree-lined promenade. Two brass bands played patriotic music. Four snowy white horses pulled the Standard's cannon through town. The four fire companies of Bodie appeared with their engines shining and their firemen decked out in new uniforms. The Booker Mine paraded its miniature cannon, "drawn by ten boys in uniform under the command of Master Charlie Irwin, who was mounted on a fine pony" (*Bodie Chronicle* 7 July 1880). People lined the streets to watch, sing, and cheer.

There was also the traditional appearance by a "collection of local clowns," known as the Horribles. They wore "grotesque masks and outrageous" dress and lampooned any- and everyone in town. Their ringleader's address, reproduced in the Bodie papers, included the following reference to the famed whiskey of Bodie and its effects:

> One poor creature accompanied by two riding and pack animals in passing through Bodie on his way to Mexico unfortunately stopped at the Senate Saloon. He hoisted in half a dozen drinks, filled a gallon jug, and continued his journey. He said to the Bishop Creek Vigilance Committee,* who were anxious to know how he came into possession of 300 head of horses between this place and Independence, that it was Bodie whiskey, and he was afraid it would choke him. He didn't lie about it, either. (*Bodie Chronicle* 7 July 1880)

In the afternoon, there was the traditional track meet featuring the hammer throw, high jump, and "hop, step and jump." The wrestling tournament offered a first prize of $100.

The fireworks display, however, fizzled: "There was scarcely enough force in the powder to carry the Roman Cannonballs above a man's hat."

---

* The Bishop Creek Vigilance Committee was a group of vigilantes patrolling the area for livestock thieves.

THE BOOM YEARS SOCIETY

Most of the games are patronized by a truly cosmopolitan crowd of customers from all the walks of life. And at the tables may be frequently seen representatives of nearly every portion of our country, as well as those of Europe. The man with a rich brogue from the Emerald Isles, the descendant of the noble ancestors, who in kilted skirt and tartan "we Wallace bled," and the natives of the great Empire on whose soil the sun never sets, the one who was born in Veterland, the Swede, the Norwegian, the child of sunny Italy, the son of La Belle France, the African, and even the almond eyed Chinaman frequently takes his chance copping the ace. . . . Bodie is a lively place and . . . contains all kinds of people. Everybody coming here can find what they want—if they are looking for a fight, they can drop into it and get gloriously whipped in three minutes of entering town. If they wish to behave themselves, they will be safe and as little interfered with as they could be in streets of New York or Boston. —Weekly Standard-News *(25 December 1878)*

In 1880, Bodie's reputation as a wild den of bad men was circulating widely. However, as you settled in, you would soon see the stabilizing factors of Bodie: the town's layout, the accepted and expected standards of conduct, the social organizations. These gave the town a social framework that its rowdier elements livened up a bit, but within the tacitly established times and places.

Bodie was a town with an unusual combination of lawlessness and safety. For example, you would never really feel the need to lock your door. Only the banks locked their doors. Private homes were not only left unlocked, most didn't even have locks to begin with. At the same time, you were living in a town with the highest murder rate in America, then or now (McGrath 1984).[3]

Bodie was well provided with State, county and town laws, fairly well enforced. There was some shooting and killing in Bodie, but almost wholly among the toughs themselves, and peaceful citizens were seldom molested. Bodie had . . . people fully the equal in brains, education and culture of those in any community of the same size anywhere. (Colcord 1928, pp. 112, 120)

Besides the usual "respectable" businesses in Bodie, you would also find more than the usual number of saloons and brothels. But then, the gold-mining camp lifestyle, with which Bodie began, included more drinking and carousing than the average American city's lifestyle.

NOT ONLY OUTLAWS HAD GUNS

Bodie was a town full of people living on the edge. Most residents, especially most men, carried guns, often concealed. The average miner did not stride about with the holsters of the western movies on his hips; most men kept their guns stuffed in the waistband of their pants or in a pocket. Gunfighters who were still alive when the first Hollywood depictions came out scoffed at the holstered image: It'd take too long to get your gun out, they claimed. And as for the macho man firing two guns at once, often from hip level, that would compromise aim too much. It's hard enough to aim one gun well when you're under fire; to try two guns at once would likely distort your aim badly enough, you'd be the one to get killed. "In gun fights, one was not usually given a second chance" (Giffen 1940).

It's important to note that if you lived in Bodie, you assumed that the people around you had guns and would use them. The papers are full of tales of conflicts resolved by one party pulling out a gun to make his point and the other party usually, but not always, retreating. This does appear to have had

Bodie children playing at "bad man" games. Courtesy the Cain Collection, Bodie State Historic Park

an effect on the kind of crime found in Bodie. FBI-style statistical analysis shows that Bodie's rates for assault, robbery, burglary, and general theft were far lower than most towns of its time, or of any time, in America. At least one study attributes this to the fact that would-be criminals could expect to be confronted with an armed, unwilling victim (McGrath 1984).

The same research suggests that Bodie's extraordinarily high homicide rate during the boom years might have been due to the deadly combination of heavy drinking and lots of guns. In the atmosphere of Bodie, it was easy for drunk people with lousy judgment to step outside to settle matters, only to end up in the hands (and boxes) of the undertaker.

THE GOOD AND BAD SIDES OF TOWN

Bodie definitely had its Wild West times with saloons and gunfights, dance halls and brothels, and, in Chinatown, opium dens. But these places and events were confined, for the most part, to the Bonanza Street section of town. Today, that part of town is long gone, the buildings consumed by fires and by remaining Bodieites in need of wood. Most of the rest of the town was home to more "proper" inhabitants, that is, families and others who did not commonly frequent the saloons or brothels and who maintained a lifestyle more in keeping with our perceptions of Victorian times.

There were more than 300 Chinese Bodieites, and they largely kept to their own part of town, in the King Street area, adjacent to Bonanza Street.

Now the area is mostly an open field where the visitor can wallow in tall grass and tall tales. The Chinese population in Bodie was overwhelmingly male; the very few Chinese women in Bodie were mostly prostitutes. This, plus the presence of the Chinese-run opium dens, may be part of why Chinatown was part of the rowdy, "bad" section of town.

Many Kuzedika Bodieites spent their days in the town center, but numerous former residents recall a Kuzedika segment of town, up on the hill above what is today the parking lot. Other groups, Mexican, Irish, French Canadian, and so on, tended to cluster socially to some extent as well, although they did not have their own sections of town. This was ac-

Kuzedika people were a standard element of Bodie town life. Courtesy the Bancroft Library, University of California at Berkeley

cepted and expected. "Folks and families tended to keep to themselves," recall several of the Bodieites still alive today.

If you were passing through the rowdier part of town, you expected to be careful. If you were a man, you kept your toes and stayed mindful of why you were there. If you were a woman or a child, you could generally pass unharmed near saloons and brothels because the code of respectful behavior toward you was strongly observed. Nonetheless, everyone had to be careful. Gunfights were indeed a common occurrence. Mrs. Helen Lafee, the granddaughter of a Bodie hotelkeeper, recalled her grandfather saying that when his wife took their children walking, she "would stand near the wall of a building during a shootout and the kids would hide under her skirts." Another Bodieite relative recalls that her grandfather did not allow his red-headed wife out of their hotel except for meals.

FOR THE COMMON GOOD:
ACCEPTED AND EXPECTED STANDARDS OF CONDUCT

The standards of behavior in Bodie were more or less agreed upon within the common areas of town. These standards, although they gave some structure and form to conduct, which humans need to communicate and get along, definitely favored the mainstream culture: the white, "proper" Euro-American residents.

People who were physically weak were treated with greater consideration than those who were more vibrant. These included the very old, the sick or injured, children, and, to a lesser extent, women (particularly "proper" white women). This applied to people of all backgrounds to some degree. It was not all right for a white person to be intentionally cruel or harmful to an elderly Kuzedika or Chinese person simply because they were not white. However, ridicule that we see today as cruel was generally acceptable among the mainstream white population in those days if the target was Chinese, Kuzedika, or any nonwhite. But if you were downright cruel, you could expect some social repercussions. And if your target fought back, he or she often gained some verbal, if not physical, public support.

One evening in 1883, some years after the boom time, a group of teenage boys harassed a Chinese boy returning home from his work, hitting him with a stone and kicking him until he handed over his money. People were outraged. The paper called for punishment of the boys and a curfew for all teenage boys, which was eventually implemented for a time.

A *Daily Bodie Standard* story on 7 August 1878 illustrates another Bodie attitude. A man who was "shooting at squirrels to check the functioning of his revolver narrowly missed a young girl who had opened her back door." The mother of the child, in understandable fury, hit the shooter on the head with a rock. A crowd gathered. When a policeman arrived to arrest the shooter, the crowd wanted to let the mother kill the shooter. The policeman had to produce his own gun to protect the shooter, out of fear that the crowd might otherwise lynch him.

Another accepted tenet of Bodie life was that a person's word was his or her bond. Verbal agreements were taken very seriously, verbal threats or insults even more so. If you said that you were going to kill someone, you could expect that person to defend himself the next time he saw you, whether or not you made any actual attempt to kill him. Even the local arm of the law and justice respected this code. If a killer could prove that his victim had threatened to kill *him,* it would be considered strong evidence that the killing was a matter of self-defense.

SOCIAL ORGANIZATIONS: THE UNOFFICIAL ORDER

Several secret societies flourished in Bodie, such as the Knights of Pythias, Independent Order of Odd Fellows, the F. & A. Masons, the Ancient Order of United Workmen, and so on. Affiliates existed to accommodate women as well, such as the Order of Rebecca. These organizations offered a system of social support and, at times, financial support, along with a fairly rigid set of rules of conduct for their members. These rules were backed up with disciplinary action in some cases, as in the case of Brothers Springstead and Langly in August 1880. The Odd Fellows records of that month (Independent Order of Odd Fellows of the State of California, Bodie Lodge Odd Fellows Records 1876–1916) include notation to certain members that they are "hereby notified and required to appear before the committee . . . appointed to try the charges preferred by Brother WW Springstead against Brother EH Langly at Oddfellows Hall on Aug 9th, 1880." The specific charges are as follows:

> To the . . . Officers and Members of Bodie Lodge No 279 IOOF
>
> Dear Sirs and Brothers:
>
> I herewith bring charges and prefer the same against Bro PGEH Langly . . . for conduct unbecoming an Odd Fellow, intraducing [sic] and circulating false, and malicious reports against a bro of said Bodie Lodge . . . as follows to wit:
>
> First. that between the 20th day of June 1880 and the 10th day of July 1880 at Bodie Mono County State of California the said Bro PGEH Langly stated to Bros McVey, Loddell, Holmes and Ward . . . that Bro WW Springstead . . . was in the habit at the close of each meeting of said Bodie Lodge No 279 IOOF of visiting a Public Dance House in the town of Bodie, and associating with the occupants thereof, to the extent of dancing with the inhabitants thereof till a late and unreasonable hour of the night, and then to publicly promenade the main street of said Town of Bodie at the hours of one and two o'clock in the morning with a common prostitute on each arm.
>
> Second—that said Bro PGEH Langly will know the said statements so made by him, as aforesaid, to the said Bros. above mentioned, to be willfully and maliciously false.
>
> Third—that between the dates above mentioned . . . I charge said Bro PGEH Langly with making the same willfull, malicious and false statement to Bro. TA Stephens.

The fourth, fifth, and sixth charges are basically reiterations and embellishments of the same charges. It would appear that if you were fond of passing on bawdy gossip about another "brother," you had better be certain it was true.

The societies' activities created associations among townsfolk, and provided a way for newly arrived people to get acquainted. When a member of the Odd Fellows arrived from another town, he would generally bring with him a letter of introduction from his previous lodge. The Bodie Odd Fellows, in turn, would inform his previous chapter that their fellow had surfaced in Bodie. Organization records show that their members were primarily, though not exclusively, from the white mainstream culture.

Many of these organizations offered some financial help in the case of death or disability. The records are full of notices going into and out of Bodie regarding various members injured or killed in town or in their new locations when they had moved on. For example, one note from Goldfield, Nevada, addressed to the Bodie chapter of the Odd Fellows, says,

> Dear Sir:
>
> I am writing you a few lines to inform you that George Cremer while sawing wood on the 31st of Dec accidentally ran his hand in the machine completely severing the thumb and very nearly his whole right hand. If you would kindly notify his home lodge.
>
> You would genially oblige,
> J. A. Eriksen
> PS: George says he will write as soon as he is proficient with his left hand.

The wife of a workman (or other next of kin) would receive small financial benefits from her husband's societies should something happen to him. Some organizations excluded several professions from membership. The Ancient Order of Workmen would not consider admitting a man who was "selling liquor as a beverage, manufacturer of . . . explosives, sailor, soldier, pro baseball or football player, circus performer, or submarine divers."

This was taken quite seriously. The records show that if you were an established member and you changed your profession to be one of these (in Bodie, selling liquor was naturally the most likely), you were immediately given the choice of changing jobs or leaving the organization.

Records also show that, after much discussion, it was decided that selling liquor wholesale was allowable, but the retail selling of individual drinks was not. Therefore, the wholesaling businessman could belong, but the saloonkeeper could not. This, then, helped form a sort of social hierarchy that considered selling alcohol in bulk acceptable, but selling alcoholic drinks less respectable. The organization did not keep its members from patronizing the businesses that sold the alcohol; it just separated the proprietors socially. In those times when the temperance movement was afoot and the acceptance of alcohol at social gatherings was not universal, making these small distinctions was more important than it is generally today in America.

HIGH-DESERT DISSIPATION

For the Bodie menfolk, the town was much fun when it wasn't hard work, except perhaps for the winter weather. Drinking was tolerated, although imbibers were expected to confine themselves to the part of town that featured the saloons and other features of the rowdy West.

The following article in the 27 March 1880 *Bodie Morning News* describes the possibilities for a standard Bodie evening (for men):

MAIN STREET AT NIGHT
A Visit to a Saloon, Faro-room and the Chinese Quarters—
Different Symptoms of Human Nature

Sentimental people say there are no attractions in Bodie of a night; that all one can do is to walk down Main Street to the Bulwer-Standard mill and then back on the other side to the post office and then go to bed. Bodie of a night is a fine field for one who can study the different phases of human nature. Start out, for instance, at 8 P.M. from the post office. The first thing you will strike will probably be the steps of the Miner's Union Hall. Cross Green Street, and the next thing to attract your attention and hand will be either a trout or a chicken hanging outside the Champion chopstand. If you are not at once arrested in your tour by an officer, Gunn's saloon will be reached without any exciting incident. Go inside and you will see a homogeneous

Bodie's famed rowdy nightlife featured gambling and other forms of revelry.

class of citizens at the bar, all in a jolly mood, with the exception, perhaps of one or two who have got beyond that stage and are trying to sleep sitting on the edge of Steve Goodvan's water-barrel or on Gunn's safe. Go into the faro-room. Four woodchoppers, looking on at the game and resting their dirty hands against the wall; the game went through them, but the fascination of faro is so great that they still wait in the room—and probably will till the game closes. Keeping cases is a young barkeeper of a down-town saloon, who when he wins a bet always "pinches" it, and for doing so is pinched on the arm by the man who staked him, and who stands just behind. There are a number of players at the table. Some of them, when they win on the turn, will stack up their checks in front of them and seem utterly oblivious of the fact that the deal is still going on, for just when the dealer is about to make a turn, they will reach over from the eightspot and nearly break an arm in trying to place three checks on the king, all of which provokes anathemas against them by those players who are of a nervous disposition. There are some who are down to their last white check, and a gloom is seen spread over their countenances; others have a "stack" or so of reds in front of them. In fact, nearly all the stages of wealth on a small scale can be seen at a faro table.

Well, go out of the faro-room and proceed toward the saloon door. Before you get to it, a very large man with an honest, good-natured countenance enters the saloon; he opens both the swinging street doors. He looks at the crowd at the bar and after asking: "Is this a single-out or a three-one," will invite every one in the saloon to drink. All fall in, and the visitor throws down a twenty. He puts the change in his pocket and saunters with a swinging gait into the faro-room. But, while all are drinking at the bar notice that weasel-faced man slide up to the corner of the chop-stand and put a link of sausages in his coat pocket. The eating stand waiter was included in the invitation to drink and, of course, did. Right there are the extremes of human nature: the generous man and the petty thief. Go out in the street again. You pass a cigar store. Look in. There is a miner, dinner-can in hand, who is about buying a new wooden pipe. See how he examines it: places his finger in the bowl, looks at the stem to see that there is not [sic] flaw in it and finally sucks it to see whether it will draw all right. The miner is satisfied with it and puts on the counter his half-dollar, which the salesman sweeps into the drawer with a pleasant remark at the time of "Anything else?" Move on. Peep into the telegraph office. The operator is at work sending an account of a strike in a mine that has just been written out by a Superintendent. A Deputy Sheriff is in the office. As soon as the operator has transmitted the superintendent's message the officer of the law will hand him a message of a different tenor from the one that has just been flashed over the wires to Carson—a message, the result of the delivery of which might cause a person to be incarcerated in a penitentiary. Go further down the street. There is nothing particularly interesting to note on the way to King Street. There are saloons, of course, where scenes like the one previously seen are likely going on. Stop a minute. Go into that store where articles from a paper of pins to a section of water pipe can be bought. Ask the clerk what kind of customers he has dealings with. He will tell you, all kinds. Ladies go in there to buy nick-nacks and miners call on him to purchase candlesticks. Now, go to King Street— the Chinatown of Bodie. Your ears are tortured by the sound of the Chinese fiddle and your nose offended by the thickness of Chinese stenches. Enter an opium den. There you see victims of the terrible vice of opium smoking, lying on dirty bunks, some with coats off and others divested of their boots in addition to their coats. Men of all ages are there; all have the same expressionless eyes, sickly countenance and look as if they themselves knew they were damned. Leave the filthy hole and go once more to Main Street. Visit the dance-houses. See what is going on in them; mix with the throng for curiosity's sake; see if human nature cannot be studied in there. Then cross to the other side of Main Street and take notice of everything you see. If, after your trip, which will take several hours, you are not convinced that a night in Bodie can be whirled away by observing different phases of human nature, then go to Mammoth or some other more attractive place.

Of course, as the article suggests, an entire social scene revolved around saloon life. You would soon find there was even an etiquette of sorts to the drinking ritual, and the lingo reflected the mining industry in its leisure hours. When miners were at work, the hollered "Fire in the hole!" meant head for safety; they were about to blast. In the saloons, the standard saluta-

Bodieite Ed Goodwin reports that as a boy he and a friend discovered that this now-abandoned roulette wheel was rigged.

It's the Water

If you spent much time at all in the saloons, you'd soon find that Bodieites were very proud of their whiskey. Bodie whiskey was especially famed for its power, if not its refinement:

Bodie Whiskey

A San Francisco paper contains the following base libel upon our local beverage: The barkeeper in a Market street saloon hurriedly mixed in some alcohol, kept for cleaning the mirror, and spirit of turpentine, and Jamaica ginger, and Perry Davis' pain killer, and when the Bodie man said "yes" in reply to his question whether he liked some bitters in it, shook half a gill of pepper sauce into a tumbler and pushed the bottle toward him. The Bodieite filled a heaping tumbler full and passed it off, and, when he had recovered his breath, said to the barkeeper: "Young man, that's whisky. I haven't tasted nothing like that since I left Bodie two weeks ago today. That's real genuine liquor; kinder a cross between a circular saw and a wildcat. That takes hold quick and holds on long. Just you go to Bodie and open a saloon with that whisky, and you might charge four bits a glass for it and the boys would not kick." (*Bodie Standard* 27 February 1879)

tion when a generous soul decided to buy the house a round of drinks was "Fire in the head!" Occasionally, some besotted fellow who was in the habit of buying many rounds of drinks would even earn the phrase as his nickname. Fellows called "Fire in the head" were popular in those days.

Helen Lafee grew up hearing tales of her father's childhood in Bodie: "When my dad was six years old, he wandered away from home, into a saloon and said, 'Mommie wants a drink of beer.' They obliged him with plenty and when he fell asleep, they wrapped him in the American Flag and put him on a billiard table. . . . My grandfather became very ill and it was necessary for [Uncle] Marcus to work as a bartender in a saloon when he was 14. He was tall for his age. When the two gunmen would order fantastic complicated drinks on purpose—and say, 'Kid, mix me a ——,' he expected them to kill him on the spot, but to them it was a big joke, so he survived."

They say there is nothing quite like Bodie whiskey. Courtesy the Dolan-Voss Collection

OPIUM DENS

Although the opium dens were considered to be "a Chinese thing," as one former resident put it, plenty of white men (and an occasional woman) hung out there, and at times the local papers clamored about the problem of the dens. The description of opium den ambiance is striking:

All is still, save the frying and hissing of the drug, as wad after wad is cooked and placed in the pipe. . . . The bunk is full of smoke; through its volume . . . a young man, pale as death himself, young and good-looking, and were it not for the hard seam under the eye and deathly pallor of the face, the picture of intelligence . . . the pipe has fallen from his hands. . . . See the thumb and forefinger how they work; even in his dreams they continue to twist and twirl the little wire, . . . forming another dope. (*Daily Free Press* 14 March 1883)

In general, it seems that a man could behave pretty much as he wanted to as long as he tipped his hat to respectable ladies, was polite to children, kept his money dealings square, and didn't threaten to kill anyone. Outside of that, he could drink himself under the table every night and not only be accepted, he'd have to make room for plenty of company under that table.

All in all, the hardworking newcomer could expect to find a rigorous but comfortable existence up home in Bodie.

Thy food, thy clothing, thy convenience of habitation;
thy protection from the injuries, thy enjoyments of the
comforts and the pleasures of life: all these thou owest
to the assistance of others, and couldst not enjoy but in
the bands of society.

—ROBERT DODSLEY (1817)

Bodie's Social Ladder

The social structure in Bodie, like that in most of mainstream America, placed white people on the top rung and parceled out the others below. And within each group of people, men and women were viewed and treated differently.

THE BODIE BOYS' CLUB

The white men of Bodie were the power holders, just as in other American towns of that time. They ran the mines, the law enforcement, the social clubs, and many, though not all, of the businesses. Some have already been mentioned; others came to prominence later on. They worked hard for their successes in this harsh, remote town. Many of their legacies remain today, at least in part. A few of these Bodie movers and shakers deserve special mention.

Lester E. Bell

Lester E. Bell came down to California from Canada with his good friend J. S. Cain. The Bell and Cain friendship was solid enough to make them family. Not only did they emigrate together, they also married sisters, each becoming the other's brother-in-law. Mrs. Lester Bell's maiden name was Charity "Cherry" Wells; her sister was Martha Delilah Wells.

Lester worked for many years at the Standard Mill and was well thought of all over town. J. S. Cain hired Lester on first as an assistant at his cya-

Some Bodie notables taking a spin on the ice on Bodie Bluff. A young J. S. Cain is the fourth one in from the right. Courtesy of California State Parks, 2002

Lester Bell and friend riding in the car that Lester built. Courtesy the Bell Collection

nide plant, and according to Ella Cain's account, "When Lester learned the process he took charge of the plant." Perhaps most of all, Lester was a man respected and recognized for his utmost honesty: " 'twas said no matter how much gold passed through his hands none of it ever stuck to his fingers" (Cain 1956, p. 84). This makes him stand out among men, particularly in a town as famously loose as Bodie.

The Bells' house of later years was renowned among former residents because of its long, enclosed walkway to the outhouse. This was seen as a wonderful luxury in winter back in the days before plumbing. Lester's descendants remained in Bodie long after he had passed on. In fact, Bob Bell was the last Bodieite to leave, and has been of invaluable assistance to the park staff over the years.

Harvey Boone

Harvey Boone's "Boone and Wright Store and Warehouse," built in 1879, still graces the corner of Main and Green streets in Bodie. In 1884 he bought the Gilson and Barber store at King and Main in Bodie and ran it in conjunction with his Green and Main store. In 1900, he opened a branch store in Masonic. Boone's home during the boom time was on Green Street, near what is today the rangers' office. In the 1890s, Boone ran a major freight wagon operation. Boone also owned the Boone and Wright Stables and Free Corral on Green Street and branch operations in nearby towns of Aurora and Masonic. Today, all that is left is the tiny Boone Grainery/Saddle Room (number 60 in the SHP brochure). Harvey Boone arrived in Bodie in April 1878. He came from a fairly distinguished American family. His father, William, was Daniel Boone's cousin, and his mother, Sarah Lincoln, was Abraham Lincoln's grandaunt. Harvey Boone also owned a ranch that today lies submerged under the Bridgeport reservoir.

Terrence Brodigan Sr.

Terrence Brodigan and his son, Terrence John, were in town in 1879. Terrence Sr., originally from Ireland, gained much fame in Bodie by having been linked with W. S. Bodie when Bodie first found gold in the Bodie hills. Although many accepted that he was part of that party, others denied his claim, saying that the main source supporting the idea of Brodigan's presence in the party was Brodigan himself. All acknowledge, however, that he knew Bodie from when they both lived in Sonora, California. Whether or not he was with the original Bodie party, Brodigan was a well-known character in town.

The J. S. Cain House in the 1800s. Courtesy the Cain Collection, Bodie State Historic Park

The J. S. Cain House today.

The Cain House interior was as elegant as any big-city dwelling. Courtesy the Cain Collection, Bodie State Historic Park

James Stuart Cain with an unidentified child and dog. Courtesy the Cain Collection, Bodie State Historic Park

James Stuart Cain

No discussion of Bodie is complete without mention of James Stuart "J. S." Cain. Cain, born in Quebec, Canada, in 1854, came to America with Lester Bell. He and his wife, Martha Delilah "Lila" Wells (originally of Genoa, Nevada), moved to Bodie in 1879, shortly after marrying. He started out working for G. L. Porter's lumber company and ended up owning it, the Standard Company, and most of Bodie. In 1888, he bought a half interest in the Bodie Bank, and in 1892 he bought the other half. He kept the bank operating for 40 years. He was also the Wells Fargo agent in town until the agency withdrew its service in 1912. It was the descendants of the Cains who initiated the formation of the Bodie State Historic Park, thus saving the town from likely destruction. The Cains' house is still standing in Bodie, at the corner of Green and Mills streets. In the boom time, that location featured the Bull's Head Market.

Patrick Reddy

Patrick Reddy was famous as a brilliant courtroom attorney. If you were a defendant with a hopeless case and you could retain Reddy, you could expect to see the light of day again. Although known to be very honest, Patrick Reddy played to win. His law office in Bodie was known as the "most imposing law office outside of San Francisco." Reddy was born in Rhode Island in

1839, although rumor has it he had been conceived in Ireland. He came to California as a miner in 1861. In 1864 he lost his arm after a shooting incident in Virginia City. No longer able to work the mines, he began studying the law, being admitted to the bar in 1867. He and his wife came to Bodie in 1879. He was famous for his courtroom style and amazing recall. He was just as famous for his humanitarianism and pro bono work. He favored miners' causes and, later, women's suffrage.

In 1881 he opened a law office in San Francisco and for two years kept both the Bodie and San Francisco offices open. This is quite a feat when one considers that the commute between the two places took about 36 to 40 hours, and the price was $30 to $36 each way (Wasson 1878, p. 4; Wedertz 1969, p. 23). The roads were described as "generally good, with corresponding accommodations," but by today's shock-absorbing, paved-road standards, they were a bumpy ride. Eventually he moved to San Francisco, where he died in 1900. Pat Reddy's house is still standing in Bodie.

LADIES DON'T SPIT ON THE SIDEWALK

Bodie was always an overwhelmingly male town. Even in the boom time, women made up only about 10 percent of the total population. Within this small number were respected and disrespected women.

"Respected" white women, that is, those who were not prostitutes and conducted themselves in what was considered a "proper" manner, enjoyed

Bodie ladies having a wintry chat. Courtesy the Bell Collection

In Bodie, Mother's work was never done. Courtesy the Bell Collection

better treatment on the street than did most nonwhite women. Even so, they were not welcome everywhere. The saloons were the domain of men alone, and the only women in the brothels were those who worked there.

If you were a woman, you had constraints on your behavior that safeguarded your reputation. If you were like most women, you wanted to be a "proper" lady so you might one day marry a "proper" gentleman, who would cheerfully provide you with a nice house, decent income, and wonderful children. You then labored to keep your social status as high as possible through keeping a clean house, raising well-behaved children, and putting on social functions that fit certain narrow guidelines encouraging you to duplicate East Coast urban and European styles in decor, food, attire, and music. This lifestyle was never completely secure, however, unless you had your own independent financial security. If you were left a widow—not an uncommon experience in the late nineteenth century—you had to scramble for income for your family within a very limited range of employment choices. More than one frontier prostitute entered the world's oldest profession because she was widowed with hungry mouths to feed.

If you were a "proper" woman, you did not enter saloons, you did not drink often or to excess, you did not smoke, color your hair, or spend time alone with men other than your husband. Although you were safe walking the streets alone at night (provided you steered clear of the saloons and their well-oiled clientele), you were not automatically above suspicion, recalls Mary H. Salmon, a former Bodieite. Her grandmother, a midwife, was returning home one night after delivering a baby, her nursing tools and a bedpan under one arm. A policeman stopped her, considering the bulges under her cape suspicious. "Now in those days no self-respecting lady of good moral character would admit to a man that there were such things as bedpans and in her mind she was not about to start admitting it now," Ms. Salmon explains. When the policeman inquired as to what she had, she stood tall and told him she would be happy to put it around his neck, then began to stride home. The policeman was so taken aback he let her go, but followed her all the way home.

In return for your restraint in your personal behavior, others would treat you, a "proper" lady, with respect on the street. In general, rude treatment of a lady in public by a man was frowned upon, and "using abusing language" was a punishable offense (*Bodie Morning News* 24 July 1880). A "proper" lady could join social organizations for women.

Annie Donnelly

Annie Donnelly was the wife of butcher Charles Donnelly, Eli Johl's business partner. She began Bodie's school by teaching in her own home before the schoolhouse was built. Annie stands out in town lore as a stereotypical "proper" lady of Bodie partly because of her antipathy toward Lottie Johl, Eli Johl's wife (who was a former prostitute). Annie is said to have never accepted Lottie socially, because of Lottie's shady past. Annie Donnelly was also known for her painting hobby, as was Lottie Johl. Paintings by both Annie Donnelly and Lottie Johl are displayed in the Bodie Museum today. The Donnelly house remains standing in Bodie.

Annie Miller

Annie Currie Miller was born in Minnesota and came west to Lee Vining with her family as a child. She came to Bodie from Lee Vining when she married William Miller, a miner. She ran the Occidental Hotel at one point, and later a boardinghouse on Green Street. Her good cooking was famous in the town, and the many bachelor miners were grateful for the delicious dinners she served up nightly. She was also known for preparing midnight suppers for people attending the dances in Bodie. Annie Miller's boardinghouse still stands, up on Green Street.

The Demimonde

Bodie naturally had its population of prostitutes, some of them quite famous. The enormous man-to-woman ratio made for a lively prostitution business. The popular image of gussied-up prostitutes hanging around the saloon was not the case in Bodie. The saloons were almost exclusively male. Dance halls featured "hurdy-gurdy girls" who danced with men for a fee, but generally did no more. Being a dance hall girl was not an ideal profession for a "proper" young lady, but it was not the same as being a prostitute.

Prostitutes stayed in their brothels or "boardinghouses" and did their entertaining/business there. Respected women remained at home. When "proper" men with wives and children frequented brothels, it was not lauded but greeted with a "boys will be boys" sort of attitude. The newspapers occasionally mention a particularly lively night in that part of town, which would bring "heartache to many a wife" in Bodie.

If you were a prostitute, you spent most of your days inside the brothel where you worked, with the windows closed and the curtains drawn. If you were young and pretty, you could make good money. However, your job security was worse than that of a fashion model in the twenty-first century. As you aged, you would bring in less and less money and be treated with less and less consideration by the clientele. Those who were lucky found ways to bridge the gap into respectability through marriage or career change, but they were few. Others became madams and ran their own establishments, but again, they were unusual.

Most simply slid further down the scale of respectability, ending used up, working out of tiny dilapidated "cribs" or on lonely, dangerous street corners. Others died from venereal disease, clumsily handled abortions, childbirth, drug addiction, or suicide. Still others died at the hands of violent customers or acquaintances. Many prostitutes became opium addicts, and opiate overdose was a common suicide route for prostitutes and others alike. The life of a Wild West prostitute was neither glamorous nor exciting.

If you were a "disrespected" woman, which usually meant a prostitute, you were not treated as well as the "proper" women, nor were you welcome in the greater social scene, despite the setting of a rough-and-tumble mining camp. These boundaries were taken quite seriously, to the point that when a prostitute, or any other perceived renegade, died, they were not allowed to decompose in the same ground as the proper people. The "proper" folk were buried within the fenced, consecrated cemetery. The rest were buried outside the fence, on what is fondly called "Boot Hill." Although it might seem rather strange that the arbitrary erection of a metal fence on a dry desert hill could stir so much concern, the issue of where one was buried was important in Bodie.

Prostitutes were so isolated in their section of town that some people in Bodie had never even seen one. Mary Salmon recalled hearing about her uncles' curiosity about the "Bad Men and Ladies" of Bodie. They had never seen either, so they waited until the next burial in "Boot Hill," outside the proper cemetery, happened, and when the mourners left, they scurried up and tried to dig up the body to get a look. They were quickly noticed and brought back without reaching their goal. Mary Salmon writes,

They were dragged home in disgrace to their horrified mother. In the following weeks, every floor, wall, windows, etc. were scoured and polished by the boys. To further be redeemed, each had to say a rosary out loud every night for a month, which really killed them.

My grandmother would not let my grandfather have anything to do with their punishment, knowing he thought it very funny and wasted no time telling his cronies about it, much to his wife's distress and embarrassment. Unknown to his wife, several weeks later he took both boys down to the red light district, showing them all the Bad Men and Ladies they wanted to see, and gave them all the information they wanted. If his wife had gotten wind of this foray, at that time, the poor man would probably have spent years on his knees saying rosaries.

As wives and families arrived in number, they brought with them social functions and organizations that excluded prostitutes, firming the boundaries between respected and disrespected citizens.

Naughty Women in Bodie

Prostitutes' interaction with men was not limited to sexual services. They often spent time with miners doing traditionally "female" things, such as mending socks, helping them when they were sick, or just talking. It is not terribly surprising, then, to find that some men grew emotionally attached to their favorite "girl." It is important to realize that prostitution was *not* illegal in California until after 1910.[1] These women were not living outside the law, just outside the mainstream "proper" lifestyle, as were many of their clientele.

Eleanor Dumont (a.k.a. "Madame Moustache")

Eleanor Dumont is rumored to have been born in France. She is first noted as arriving in San Francisco in 1854, where she established herself as a gambler, favoring the game of *vingt-et-un* (21). Women gamblers were unusual in small towns, though much more common in San Francisco. There she was known for her petite stature, elegant dignity, diplomacy in stressful gambling situations, and aloof manner. Henry Wells (of Wells Fargo) said of her early days, "notwithstanding her strange mode of life, her chastity was unquestioned." As she aged, she grew less popular as a charming diversion at the gaming table. Like other women gamblers (and like prostitutes), she was forced to move away from the city to the outlying mining camps, perhaps falling victim to vices common to them. It was in the mining camps she earned the nickname "Madame Moustache," because of the visible facial hair on her upper lip. She was known to have resorted eventually to prostitution to earn money when her gambling earnings were not enough. She arrived in Bodie in 1878, which aroused some interest because of her reputation in the mining camp world. She was quite dissipated by that time, and did not fare well at the gambling tables. Eleanor Dumont was found dead about two miles out of town, on the Bridgeport road, on 8 September 1879. A vial next to her body was interpreted to indicate it was probably suicide.

Lottie and Eli Johl

Lottie Johl came to Bodie as a red-light-district entrepreneur. The bachelor butcher, Eli Johl, fell in love with her, and eventually they were married. Although she kept her behavior very proper after their marriage, the "proper" society of Bodie never accepted her. The Johls' life together was

Eli and Lottie Johl's house as it stands today in Bodie State Historic Park. Photograph by Craig Hubbard

somewhat lonely. Admirably, her husband stood steadfastly by her despite repeated rebuffs to his wife's goodhearted attempts to get along in the mainstream society. Lottie took up painting as a hobby, but this irked Annie Donnelly, the wife of Eli Johl's partner, who fancied herself a painter. There are sad tales of cruel rejection, including a story of Lottie winning the prize at a masked ball only to be shunned when unmasked, but it is uncertain if they are true or exaggerated. When Lottie died, the result of an incorrectly filled prescription (another controversy),[2] her husband demanded that she be buried in the consecrated cemetery. The townsfolk resisted the idea of allowing her to be buried in the fenced cemetery, but finally capitulated with the stipulation she must be buried right next to the fence at the edge of the cemetery. Her husband accepted this.

On Memorial Day, when the townsfolk traditionally visited the graveyard to pay respects to departed loved ones, Eli Johl decorated the grave of his love and sat by her all day. The depth of his devotion is said to have stirred the townspeople, and they are said to have responded with respect for him. Eli Johl eventually left Bodie, and his final whereabouts are unknown.

Rosa May

Rosa May was a "demimondaine" of legend in Bodie. She was the fabled "prostitute with a heart of gold," it seems. She was known by all for being kind and ladylike. Mrs. Lauretta Gray, Annie Miller's daughter, said she remembered serving Rosa May in her mother's dining room: "Everybody liked her . . . she was a beautiful woman. . . . You'd never take her for a prostitute."

Although proven facts of her origins are hard to come by, it is generally agreed that Rosa May, like Eleanor Dumont, was originally from France, and she had lived in Virginia City and Carson City before coming to Bodie. In Bodie, during an epidemic, Rosa May nursed sick miners back to health until she herself succumbed. Ella Cain, in *The Story of Bodie* (1956), claimed that Rosa May had rival lovers in Bodie, both of whom outlived her and formed a friendship out of their mutual grief. Rosa May was buried outside the cemetery fence, on Boot Hill. A small monument to her stands to this day, although it is doubtful that it stands over her actual grave.

ASSAULT AND BATTERY IN BODIE

The rough-and-tumble activities in Bodie were not always limited to men. Brawling women, surprisingly common, were viewed as amusing, though not proper. Generally, battles between prostitutes were over stolen business or personal items, and often at least one of the combatants was extremely drunk. Several newspaper reports portray skirmishes between prostitutes as a great form of entertainment. Even the more unusual case of "proper" women mixing it up was written up in the newspapers as an amusement: "Pay days and fighting women make times interesting" (*Daily Free Press* 6 December 1879).

Men beating women was not unknown, either. Judging by newspaper reports, such beatings were not considered acceptable, but they were not the cause of horror they are today, provided no one was severely hurt. Men beating their wives was less common than men beating prostitutes or other women outside of their families. Sadly, the prevailing attitude regarding domestic abuse was that although it was too bad these beatings took place, they were not anomalies and at times provoked jokes.

Women who defended themselves against their assailants were generally considered reasonable in their actions, whether they used fists, guns, or anything else at hand. Not uncommonly, the men who assaulted them were arrested and prosecuted. Likewise, in the few reported occasions of women beating men, the women were arrested. However, this could mean a jury trial, and Bodie juries were notorious for acquitting far more often than convicting.

Then, as now, more than one abused wife was finally murdered by her husband. When a former prostitute named Ellen Fair was beaten to death by her heavy-drinking miner husband, Job Draper, he was dispatched to the penitentiary. Their final argument had been overheard by two passing wood-

The Woman Beater

There appears to be quite a mania among a certain class of cowardly brutes, whose presence Bodie is cursed with, to beat and abuse defenseless women. Scarcely a day passes that there is not a case of this kind. The last, the beating of Mrs. A. Wylie, a respectable woman who owns and keeps a boarding house, caps the climax. One John Harrington is the brave man guilty of this contemptible action. If it is true, as stated to us, he deserves the tar-bucket and the rail, with a preemptory ticket to leave. A man so low as to whip a woman is not fit to live, much less to remain in this community, and, we hope, when shown to be guilty, he will be driven forth like a vicious dog. (*Bodie Morning News* 18 March 1880)

cutters one night, and her body was found by another woodcutter the next morning. Draper at first claimed she had fallen, but when the two witnesses came forth, his alibi quickly faded. Newspapers condemned him roundly.

Some cases fell into a murky zone of punishment not always fitting the crime. Several years later, when Minnie Williams was found dead, and a coroner's jury determined that she had been killed "by abuse and blows received at the hands of her husband Mike Williams," the records show that originally Mike Williams entered a guilty plea, paid a $60 fine, and was released. These proceedings are crossed out, and it is noted that a few days later the case was sent to the Superior Court in Bridgeport. There, Williams was held without bail after two Kuzedika men, Natchez and Captain John, testified about the matter.[3]

It seems likely that an abused woman, respectable or not, had very few places to turn, a fact that probably contributed to the abnormally high suicide rate among Bodie women.[4]

OUT OF THE MAINSTREAM

In boom-time Bodie, the majority culture was white European-American and European immigrant: "With the exception of 350 or so Chinese, a few dozen Indians, and 19 'colored,' Bodie was all white" (McGrath 1984).

There were people of many different ethnicities in Bodie (as the newspaper article quoted at the start of "Boom Year Society" illustrates),[5] and their traditional holidays were all celebrated. As stated earlier, the celebrations were not always limited to people of the specific culture. On 16 September 1879, the Mexican Bodieites celebrated the seventy-first anniversary of Mexico's independence from Spain by firing guns and flying the Mexican flag from a pole they erected at Main and Green streets. They had a parade starting from the Miners' Union Hall, parties at various homes in the town, and a dance in the evening. The *Daily Bodie Standard* noted that day, "It is always the custom of the Mexicans, through courtesy to the American people, to invite an American to act as deputy to the Grand Marshal. The honor was conferred today upon our esteemed citizen, Mr. John C. Turner, who filled the position with dignity and grace."

A few days later, Rosh Hashanah was noted in the same paper: "The Hebrews of Bodie celebrate their New Year today and tomorrow at the Masonic Hall. They have no Rabbi, but the services will be conducted by D. Rosenthal" (*Daily Bodie Standard* 16 September 1879).

The nineteenth-century Bodieites did not have the cultural sensibilities of Americans at the turn of the twenty-first century. But there was a certain acknowledgment of the different ways of different cultures, especially the Chinese and Kuzedika, and in the sections of Bodie where the Chinese and Kuzedika lived, respectively, the customs and ways of their specific cultures prevailed.

When things went wrong, everyone was entitled to his or her day in court, and there are plenty of judicial records proving that "white" was not *always* "right" in Bodie (although prejudice did seep into the courtrooms at times): "Chinese and Mexicans seem to have been treated no differently by the legal system than were other . . . Bodieites" (McGrath 1984, p. 253). At the same time, many Chinese people preferred to avoid the U.S. judicial system when possible, keeping their business confined to Chinatown and the Chinese secret societies that handled things there: "The authorities quickly learned that if both the victim and the perpetrator of a crime were Chinese, they could expect little or no cooperation, even from the victim himself" (McGrath 1984, p. 253).

Despite the legal consideration nonwhite Bodieites could expect to receive, they did have to put up with the prevailing attitudes of the time, which tended toward the assumption that the majority culture's way of life was superior and others were fair game for ridicule. Most Europeans in Bodie fit in easily: The majority culture was derived from European style and values. But the Kuzedika and the Chinese were in different situations.

THE KUZEDIKA

The Kuzedika *were* Americans: Today most people feel that American Indians are the *original* Americans. But in the time of Bodie's beginnings, the incoming settlers saw them not as original inhabitants or owners, but as people to be pushed aside and quelled. There was also more than a little fear of them because of recent bloodshed between various American Indian tribes and white settlers. Many settlers were not particularly concerned about differentiating between tribes of American Indians. A deep and rather ugly prejudice toward all American Indians tended to take over, with a specific attitude of disdain for desert-dwelling Indian tribes, all of them lumped together under the term "Diggers."

If you were a Kuzedika Bodieite, you generally lived outside of the Euro-American "proper" section of town. You grew up speaking the Kuzedika language and learned English as a second language. You and your family spoke English differently than the white settlers. Although you and the white people could understand each other, it was at times a little awkward.

The elders of your tribe were the teachers and the decision makers for the

A Kuzedika round house in winter.
Courtesy the Cain Collection, Bodie State
Historic Park

A Kuzedika woman with her baskets.
Courtesy the Cain Collection,
Bodie State Historic Park

group. When they were growing up, the whole area had been their home. The tribe's main concerns had dealt with gathering food, preparing for winter, and staying aware of other tribes' activities. Their approach to life sustained itself year to year by careful consumption and use of the land and its features. Each task had its proper place, rhythm, and ritual.

Since the settlers' arrival, however, and the enormous bonanza of Bodie, these lands had been taken by people who arrived with weaponry that surpassed anything your tribe's elders had ever seen before. The traditional Kuzedika lifestyle was no longer allowed. You could not hunt and gather on ranchers' and farmers' lands. They would threaten you or kill you if you tried. You could do jobs for the settlers and receive their money in return, but your elders were not accustomed to this. The new ways did not fit well with the established ways, but the old ways were impossible with the new social and physical constraints. Still, some traditions survived for a while.

The traditional kutsavi- (and, in alternate years, piaghi-) gathering celebration on the shores of Mono Lake continued despite the changing landscape.[6] The local press noted one year, "About 1500 Paiutes are gathered at the north east corner of Mono Lake holding their annual festival. This consists of feasting, drinking, horse racing, and going through the war dance" (Loose [1979] 1989, p. 152).

In the Bodie boom times, many Kuzedika women combed through the tailings piles, searching out the rock with enough gold still in it to redeem for money. Other Kuzedika women earned money doing odd jobs for the women in the mainstream section of town. Likewise, many Kuzedika men worked odd jobs for families around town. It is not noted if there were any Kuzedika-owned businesses in Bodie.

Interviewed Bodieites recall a Kuzedika settlement over the hill, between the graveyard and what is now the parking lot. They say that in later days, several Kuzedika lived in town. There were Kuzedika-Anglo married couples as well. Mrs. Lauretta Gray, Annie Miller's daughter, remembered that the little brown house that is left up near that hill was a Kuzedika person's house.

Esther Leeds, who was born in Bodie in 1892, wrote to the park about her time up home in Bodie. Her father was a pastor at the Methodist church, and her mother played the organ. She recalled that her parents knew several Kuzedika Bodieites: "[T]hey made many friends there. Among them were Indians who would bring herbs, etc. they thought would color my skin to match theirs. They thought I was too white and called me 'Buckskin' and that name stuck with me for a long time" (Leeds 1986).

Although it is true that the occasional statement reflecting a more balanced attitude could be found in the press of the area, the more common sentiment was the loathsome "The only good Indian is a dead Indian." The hardship of being forced out of one's well-established, traditional lifestyle, only to be repeatedly humiliated by the well-armed newcomers, is difficult to comprehend. The depth of the cruelty can be surmised from the following story about Louis Sammons, "the pioneer of Mono Lake," from the *Bodie Standard* of 21 July 1880. It seems that Sammons

sank in that extraordinary sheet of water [Mono Lake] the dead bodies of three native-born sons of the forest[7] with a view to their petrification. The other day he brought in one body to land and sawed off a portion of the scalp, which he left at the Grand Central hotel in this city. It is just as fully a piece of petrified humanity as ever was seen. The remainder of the body was again committed to the turbid waters of the lake. It is Louis' intention to let the bodies remain in the water six months longer, at which time he expects a perfect petrification of the whole of them. . . . A queer fellow is Louis, and worthy of commendation for his experiment. There is probably no better use that the Indians could be put to than immersion in Mono Lake with a view to their petrification. Thousands of people in the old world have read the glowing accounts in Cooper's[8] works of the magnificent personnel of the native sons of the forest, and but few know what miserable human brutes they are in reality. Mono Lake is capable of petrifying a million of the aborigines and we advise the immersion of a thousand squaws. Friend Sammons has struck the right business, and it is to be hoped that his first installment of Indians may lead to a large demand from European countries. If it only takes three years to petrify, an immense fortune is in store for a company who will follow up the business.

THE CHINESE

The Chinese residents of Bodie were not in America with the intention of settling in for the long haul. Most of the Chinese who arrived in California during Gold Rush times, as many Europeans also, had come to earn money and return home flushed with success and riches. Of course, it did not turn out that way for many of them.[9]

The desire and compulsion to return to China were so deeply ingrained in Chinese nationals that when a Chinese person died in Bodie, his or her interment there was considered strictly temporary. The person's bones were returned for burial in the soil of China as soon as possible.[10] This practice came from a mixture of spiritual and nationalistic loyalties. It was important to be laid to rest in the family plot in China. At that time many Chinese practiced ancestor worship; therefore, Chinese people felt the urgency of keeping in line with proper spiritual ritual. In China, you would be laid to rest with the rest of your family; your descendants would care for your grave, and you would become part of the family heritage. The remains of many of the Chinese people buried in Bodie graves have been returned to China.

Credit Ticket Arrivals

If you were a Chinese man, and almost all arriving Chinese nationals were men, coming to California in the late 1800s, you most likely came via the "credit ticket system." You bought your ticket in China on credit and signed a contract to pay back the sizable debt in cash. The ticket was good for passage to California and the necessary paperwork to enter the United States. Once in America, how you earned the money to pay back your debt was up to you. Although you did not have any time limit on repaying the debt, it had to be resolved before you could return to China.

In California, you could go to an appropriate agency in San Francisco's Chinatown to get the paperwork needed to prove you were free of debt. You could then return home. But you would not dream of trying to return home without this certificate: It would reflect so badly on you and your family as to be unacceptable. The social pressure on this issue was tremendous, and although using the credit ticket system was perfectly acceptable, returning home before the debt was repaid was not.

Once you arrived in Bodie, you headed straight for Chinatown. Naturally, you would feel more at home there; you looked, sounded, acted, and thought more like the people there than those who lived in the other parts of town. A miner from Ireland and a miner from Cincinnati might have a good chat over a shared stew, but neither would find it as easy to chat with you over your traditional Chinese dinner. Nor would it be any easier for you to settle down with a European-American over a big slab of cheese, most likely a very foreign and repulsive substance to you and other Asians.

These kinds of deep cultural differences made communication difficult, and the fact that you didn't intend to stay in this country made you less interested in embracing the emerging mainstream culture in America: Not all those mainstream values existed in your family and culture. And the Americans certainly saw no reason to change their ways. As is always the case, though, simply living side by side brought some exchange of ways and ideas and, consequently, some cooperation.

Finding Work and Getting By

You might find work hauling wood to town using mule teams, you might work in the proverbial Chinese laundry, you might find any number of odd jobs around town. But attitudes of the day dictated that you would not be a miner: Chinese people were not eligible for membership in the Bodie Miners' Union. Many other forms of labor were also forbidden, some by official means, others by informal exclusion. This is why the images of the Chinese laundryman, railroad worker, cook, and opium den owner are our associations with the Gold Rush times; those were the most common types of work Chinese men could find.

As a Chinese man, you were in an awkward position during Bodie's boom years. The overall atmosphere was one of suspicion toward Chinese people. This was true throughout the United States, but most intensely in California, the port of entry for most arriving Chinese people. The constant discussion of "the Chinese problem" in the newspapers of the West, Bodie included, paints a picture of extreme distrust and intolerance toward all Chinese people. It must have been difficult to mingle with people whose newspaper editorials demanded the immediate deportation of you and your countrymen because you were considered dirty, dishonest, and less than human.

A *Bridgeport Union* item of 21 August 1880 illustrates this attitude: "The jury in the Sam Chung murder case disagreed and a new trial will be in order. Being a Chinaman he was not so fortunate as he would have been had he been a white man." It should be noted that Sam Chung retained the famous Bodie attorney Patrick Reddy, and Chung was eventually acquitted.

Some Chinese traditions attracted the Anglo townsfolk; others repelled them. The Chinese New Year

A Bodie Chinese laundry stove, complete with irons.

Opium was perceived as "a Chinese thing," although many non-Chinese partook as well. Courtesy the Cain Collection, Bodie State Historic Park

celebrations amused and intrigued all Bodieites, as the following newspaper coverage demonstrates:

The celestials[11] wound up their festival season with a sudden and emphatic outburst of enthusiasm last evening. The oriental feeling found vent in firecrackers, bombs, and free lunches. King Street, at one time and another during the night was noisy and John[12] felt hilarious. Many were drunk with enthusiasm, others with opium, while a few tackled whisky and were thrown down. White men—opium fiends—took advantage of the occasion by getting stupid over the free drug, and it was happiness all around and in the center. (*Daily Free Press* 8 February 1881)

The traditions of Chinese funerals, with hired mourners and elaborate feasts left at the grave site, amused mainstream Bodieites. Many times non-Chinese Bodieites (most commonly teenage boys) would consume the funeral feast once the mourners left.

However, the tradition of binding young women's feet was met with horror among the mainstream Bodieites. A local paper item explained that it was also met with horror by some Chinese people as well:

Miss Norwood, a lady in the American mission at Swatow, has given some interesting particulars as to feet distortions in China. This abominable barbarism is . . . vehemently opposed by the "Hakkas" and where people of this province—who, happily for Chinese women, seem to have nomadic habits—settle it begins to disappear. Of the women attending the missionary schools in Swatow about 60 per cent have their feet bound. The binding does not take place until the child has learned to walk. The pain is intense when the process is performed in the case of an adult, and source of dreadful discomfort in any case, yet the desire to have small feet is so intense that girls will slyly tighten their own bandages in spite of the pain. (*Bridgeport Union* 29 May 1880)

According to Judy Yung, author of *Chinese Women in America* (1996), the desired foot size among practitioners was three inches (although most women could only manage somewhere between six and nine inches). In the United States, only women of the merchant class had bound feet, and they were never more than 5 percent of the population. However, that 5 percent really stood out from the others. The Chinese prostitutes did not have bound feet. Although it seems barbaric by today's standards, many Chinese people of that time considered "lotus" feet to be a sign of gentility and good breeding. Many Chinese men found them extremely appealing sexually. Similarly, the extreme waistlines produced by tight corsets of the time were considered attractive and desirable in Europe and America, despite the physical discomfort and harm they brought the women. Yung maintains that likely few if any Chinese women in Bodie had bound feet (Yung 1996).

Chinese Women in Bodie

If you were a Chinese woman, by virtue of your role within the Chinese culture you had less exposure to the gaps among cultures in Bodie. There were the same two "classes" of Chinese women in Bodie as with the other women: the "proper" women, who numbered very few and were wives of businessmen, and the "improper" prostitutes, who were the majority. Not very many "proper" Chinese women ventured to America. This stemmed from several factors: partly cost and paperwork, and partly a general concern about safety and racism in America. It should also be noted that women growing up in China of the 1800s lived by rules far different from the rules that governed women who grew up in America at the same time.

Chinese society at the time was based on "Confucian ideology," which encouraged everyone to know their place and obey their superiors. This was

especially demanded of women, who could look forward to the "Three Obediences" in their lives: first to their father; then, after marrying, to their husband; and then, if they should become widowed, to their eldest son. The
three obediences were enriched by the "Four Virtues," namely, chastity and
obedience (viewed together as one), reticence, pleasing manner, and domestic skills. It is not surprising that these guidelines produced women who
were exceptionally passive and accepting of whatever the men in their life
told them to do. It was the responsibility of those men to provide for the
women, which was often a problem for poorer families.

For families with few resources, there was an accepted tradition of selling
daughters into indentured servitude to a wealthier family. In exchange for
her work as a domestic servant, she would receive care and feeding until she
became of marriageable age, when she would be set free to marry. This was
considered an acceptable and respectable way of looking after your daughter
if you did not have sufficient money. The daughters accepted this, as society
dictated they must. A bit earlier in Euro-American history, the European tradition of indentured servitude was similarly accepted by society.

Many of the Chinese prostitutes in California were sold into servitude
in this time-honored manner of China, then resold to another owner who
imported them for prostitution. Others were promised marriage, only to
arrive and find a brothel waiting for them. A few were actually sold know-
ingly into prostitution by impoverished parents. This was horrifying to many
Americans, but the Chinese attitude toward prostitution was not as punitive
as the American attitude. Prostitutes were not treated as social outcasts, nor
was prostitution seen as a moral flaw. Prostitutes were seen as "daughters
trying to fulfill their filial duties, out of economic necessity [having] been
forced to take on that role or job" (Yung 1996).

The Chinese women who worked as prostitutes in Bodie and elsewhere in
the state were ostensibly bound by a contract, usually signed with a thumb-
print because the women were illiterate. The contract stated that the woman
would be working as a prostitute, her body the property of her owner for
the duration of the contract. Many women found that as they worked off
their debt, the debt also increased because they were docked for sick time.
Sick time included time lost during menstruation and pregnancy, as well as
illness.

There was quite a market for Chinese prostitutes in Bodie and the
rest of the state. In California, there were more than 18 Chinese men for
every Chinese woman, according to the 1880 census, and the California
Anti-Miscegenation Law prohibited Chinese men from marrying Caucasian

women.[13] In 1875, a federal law required that Asian women be screened by the U.S. consulate in China before leaving for America, to prohibit prostitutes from entering the country.[14] In 1882, the federal Page Law was enacted, limiting the immigration of Chinese women to certain classes only (for example, merchants' wives), further reducing the number of Chinese women coming to California.

If a Chinese prostitute were lucky enough to get out of prostitution, she had a decent chance of marriage. Although promiscuity was not looked on with favor, Chinese society did not see the woman who was a prostitute as being at fault in her situation and considered her an acceptable candidate for wife and mother.

This social acceptance of the Chinese prostitutes was another area where the Chinese and the Americans were in moral conflict. Because prostitutes were held in their profession by a debt that few lived long enough to pay off, the mainstream culture saw them as virtual slaves. This was considered particularly offensive at that time in America, when the devastating Civil War had been fought not much earlier, largely over the issue of slavery. In Bodie, the occasional newspaper item about a Chinese woman referred to "her owner" when describing the Chinese man who appeared to be in charge of her.

And yet, in Bodie, as everywhere, the social issues came down to people interacting with each other.

Esther Leeds recalled her parents' Chinese friends: "[A]nd there were Chinese among the friends, one especially called Toy Lee, with his jet black hair and what we called a 'pig tail' hanging down his back—he was so good to us. I can still see him with his Mandarin-type black silk embroidered shirt—his shoes were also silk with colorful embroidery—I think he stands out in my mind because he brought me fortune cookies and Chinese candy."

AND IN THE END . . .

In Bodie's final days, those years just before World War II, even though the town's population was a fraction of its boom-time figures, the town remained a settlement of white, Kuzedika, and Chinese people hacking out a living together in the high desert.

BODIE AT HER BIGGEST

> *Then it came to wicked Bodie*
> *Where the men are often brutal*
> *Very many: seldom sober*

And the women mostly naughty
Bodie not so big as Gotham
Nor so broad as flat Chicago
Nor so fine as Cincinnati
But for pure and perfect badness
And for cussedness unrivaled
Faro, Keno and draw Poker
Bodie town will take the biscuit
It will knock the ripe persimmon
Right off the topmost branches . . .

—HAROLD BRADY CARPENTER *to* FRANK GILCHREST,
in Lundy, California (23 January 1881)

"The Chinese Must Go"

Occasionally someone would rise above mob attitudes and make an original statement, as in the following newspaper item that spun irony off a negative but popular phrase of the day:

A proof of the ingenuity and perseverance of the Chinese race can be found on Mill street, opposite the Mono House. The John's [*sic*] used to dry their wash cloths on the vacant lot opposite the Standard-news office, but Captain Porter enclosed the lot, and, as no other people would think of doing, our Chinese friends fixed a sort of clothes rack on the roof of their shanty, and rigged a small tackle for hoisting baskets full of wet clothes to the drying place. It is amusing to witness the modus operandi of these industrious people. The Chinese must go —[ahead]—and make money when their American brethren are loafing about the streets seeking for work and praying they may not find it. (*Bodie Morning News* 22 July 1880)

1880: THE CENSUS YEAR

As with many aspects of Bodie history, there is disagreement about just how many people lived there in the peak time. Some sources report Bodie's 1880 population as 7,000 inhabitants (Loose [1979] 1989); others place the popu-

Waiting for the Wells Fargo Express wagon (on the photographer's left). Courtesy the Bancroft Library, University of California at Berkeley

Waiting for the Wells Fargo Express wagon (on the photographer's right). Courtesy the Bancroft Library, University of California at Berkeley

lation higher, closer to 10,000 (Bodie SHP Brochure) or even 12,000 (Cain 1956). The year 1880 was a federal census year, and Bodie was surveyed along with every town, city, and farm in America.

The 1880 federal census records are far from accurate. The census itself notes that in areas such as Bodie, "[t]he list is not by any means complete. . . . The figures can be considered as only approximate, as the limits of such places are not sharply defined." The census lists Bodie's population as 2,712. It also notes the combined populations of Bodie, Bridgeport, and Mill Creek as 6,001, and we know Bodie was the largest of these towns. It has been acknowledged that although the newspapers pleaded with the citizens to be helpful with the census takers, the census was very poorly carried out. The census takers did not take into account miners living in Bodie boarding-houses. Census takers, who had fewer job prerequisites than a mannequin, considered the boarders transients, despite the fact that a large number of Bodieites lived this way for months or years at a time. Still others in town were inclined to avoid the census takers, not wanting to be recorded. Bodie was the kind of town where, if you wanted to be left alone and you behaved yourself, you were left alone.

If one estimates all the miners considered nonresidents, all those who avoided the census takers, and those whom census takers considered invalid because of their ethnicity (large numbers of Chinese, Indians, Mexicans, and others were not noted), 10,000 remains a likely figure for the 1880 population of Bodie (Scanavino 1968; Shipley 1995). Steven Scanavino[15] said of the census takers, "[T]hey decided to just settle for the people that had permanent jobs, and business people with homes. . . . [T]he rest of them went unlisted as not being permanent residents, which was true because every day some turned up missing, shot down, hauled to the cemetery."

WINTER DOLDRUMS

The 1879–1880 Bodie winter made the mining business particularly difficult. January was harsh, with lots of snow and a temperature of -26°F at the end of the month. The firewood supplies ran low. Wood packers tried to step up their activities to match the need, but conditions were rough. The wood roads were blocked with snow. One packer had three mules freeze to death on the Bridgeport road. By April, most mining companies and many citizens were out of wood. Several mines shut down. Later that month, the wood dealers finally paid for a road to be cut through the snow, enabling them once again to deliver wood into Bodie.

Excitement and Eccentricities

The constant tension and excitement over gold discoveries and mining
at times led to scenes of absurdity. One fine June evening Dr. Blackwood
decided to "jump" a claim on Standard Avenue. Bodie residents were flab-
bergasted early the next morning when they discovered that part of the street
itself had been fenced and a shack hastily built on the fenced-in claim. The
general consensus was that Dr. Blackwood had overstepped his bounds. The
fence and shack were dismantled rather quickly, the only lingering remnant
being various discussions in the newspapers. This was not to be the last peculiar behavior out of the good doctor.

STEAMBOATS AND RAILROADS

In May 1880, the Bodie Railway and Lumber Company acquired and assem-
bled the steamboat *Rocket,* initiating cruises on Mono Lake in June. The lumber company had bought the *Rocket* to transport wood and lumber across
Mono Lake from the southern to the northern shore, near the Cottonwood
Canyon road into Bodie. Wagons transported the goods into town from
there. The lumber company later decided to build Mono Mills, a lumber sawmill near the timbered area by Mono Lake's southeastern shore, and a railroad linking the mills and Bodie Bluff. During the construction, the *Rocket*
transported various goods to the shore near the mill site.

The *Rocket* became popular as a pleasure cruise vessel for a short while,
but Mono Lake was, and is, unpredictable. It can be enticingly calm minutes before the wind rises, when it turns deadly. J. S. Cain discovered this
when he was captain of the *Rocket* on one particularly harrowing voyage. The
rough conditions forced him, with his boat full of people, horses, wagons,
and equipment, into an unscheduled stop. They spent the night in haystacks
and resumed their trip in the morning. That trip permanently quelled Cain's
enthusiasm for lake transport (Billeb [1968] 1986).

JUNE SNOW AND BODIE BAD MEN

June 1880 featured slight snow and icicles, but of even more interest were
the stagecoach holdups, which were emerging as a regular occurrence along
the road between Bodie and Hawthorne, Nevada, many miles to the east.

Overall, the level of reported violence became much higher than in previous years, setting off a reaction among the "proper" townsfolk. One of the newspapers even suggested instituting chain gangs in Bodie, although not much came of it.

Many citizens were distressed by the fact that the police or "constables" disappeared after dark, leaving the town vulnerable to its rougher elements. The *Daily Bodie Standard* went so far as to offer a reward to any ruffian who could actually *find* a constable on the street after dark! The police responded angrily to the paper's editor, who replied that his offer was only intended to get the police and the criminals "in close contact with each other, as there appears no other way of getting them together." Within the next month, the citizens had removed a few of the constables and hired some new blood. Soon everyone felt slightly safer on the streets of Bodie at night.

BODIE'S FIRST CHURCH AND NEW JAIL

In August 1880, the *Weekly Standard-News* noted that the Chinese Bodieites were the first citizens to have a house of worship when they bought and converted the old Sonora Dance House into a joss temple.[16] The *News* also noted that the Catholic and Protestant Bodieites were still "in the fund raising stages" for their houses of worship. Stuart and Sadie Cain remembered, along with the laundries, stores, and herbs in Chinatown, the joss temple with its vessels of oil. They said non-Chinese children liked to blow out the flames in the temple lamps, which angered the Chinese people.

In the same month, the paper announced the construction of a new jail, nicknamed the Hotel de Kirgan, after J. Kirgan, the jailer. The new jail held sixteen people comfortably (so to speak) and included a twelve-foot fence around an exercise yard. Later, the yard was seen as less wise after two exercising prisoners laid a plank against the fence, climbed it, and slipped off on marathon-running practice.

Bodie's population peaked in late 1880. More than 1,800 houses had been built, along with numerous businesses, mills, and hoisting works. The construction work in Bodie then leveled off. Bodie's mines were beginning to feel pinched for a lack of new bonanzas. Correspondingly, there was not as much demand for lumber.

Because many mining clients could not pay their bills to him, G. L. Porter, founder of the Bodie Railway and Lumber Company, found himself unable to make ends meet. In August 1880, his company's creditors decided to place the lumber company operations in the hands of two merchants, A. F. Bryant and a Mr. Abner, and one employee, J. S. Cain, by making them signees of the business. Early the next year, J. S. Cain and a Mr. Stewart entered into partnership and became agents for the Bodie Railway and Lumber Company. They were soon "the principal lumber and wood dealers in Bodie" (Wedertz 1969).

Although things were looking up for such Bodieites as Mr. Cain, the mining business was winding down for more. The slow emigration out of Bodie began, barely after the frenzied influx had ended. The Bodie papers' first reactions were gleeful, because the first to leave were those perceived to be troublemakers.

Nearby settlements noticed the exodus from Bodie as well:

Our sister town, Bodie, has been, and is being, favored with an exodus of hundreds of the undesirable portion of its population—mostly of the rougher class that mining towns are usually cursed with. That this exodus will be a benefit to the town is not to be questioned. There have been too many people there for the amount of work doing in its mines. . . . The next pay days, we believe, will find Bodie better off financially . . . making happy homes and adding to the well-being and prosperity of Mono's great metropolis. (*Bridgeport Union* 21 August 1880)

It is not surprising that those who chose to live a life of excitement on the fringe would choose to leave when the party was starting to wind down and cash was becoming tighter:

There is no person living in Bodie for his health or to attend a picnic. There is constant call on every side for money, and the more that is secured the better. . . . Among the many resolves made to day will be one to the effect that "I shall not spend another winter in Bodie"; but like all the others it will be broken. (*Daily Free Press* 1 January 1881)

*Permanency and extraordinary dividends
are incompatible.*

— J. ROSS BROWNE (1869)

The Bodie Fever Breaks: 1881

Hopes for continued growth and riches in Bodie were running high as 1881 began. Then the San Francisco Stock and Exchange Board suddenly went limp, and Bodie stocks with it. Some fair-weather citizens moved on, once the roads were passable. Although it was the entire stock market that stumbled, Bodie, like other mining camps, was susceptible to rumor and popular opinion. With the lackluster economic situation, it became harder to attract outside capital to further develop the mines, and the locals didn't have much extra cash to spend, either.

In January 1881, the *Daily Free Press* declared that Bodie stocks continued to "attract attention" on the San Francisco Stock and Exchange Board, and that it was time to consider a water system to meet the needs of a large population because it was obvious that Bodie would be around for a long time. Although the paper was putting on a good face, doubts began to stir in town.

About this time, the issue of Chinese immigration began to heat up at the national level. Bodieites were very interested. The entire Pacific Coast was riled up over the continual influx of Chinese, and although many mainstream citizens in Bodie were against the immigration, not all found it an issue worthy of passion or violence. Newspaper accounts note much higher levels of anti-Chinese activities elsewhere in the state than in Bodie.

Bodie at its biggest. Courtesy the Gray-Tracy Collection

THE TRELOAR-DEROCHE AFFAIR

January 1881 also saw a solemnly noteworthy event: a passionate murder and Bodie's only documented vigilante lynching. The cold facts of the case, as presented by the various newspapers and documents of the day, are as follows: A French Canadian carpenter, Joseph DeRoche, shot an English miner, Thomas Treloar, at point-blank range in the back of the head at 1:30 in the morning on Friday, 14 January, on Lowe Street at Main (today an empty spot near the county barn). The presumed reason was conflict over a woman, Treloar's wife, Johanna, and possibly insurance money as well.

Johanna was working at a ball that night, and DeRoche asked her to dance. As they danced, Treloar entered the ball, saw them, and became angry. As he left, DeRoche followed him. When Treloar paused down the street to light a cigarette, DeRoche shot him. Unbeknownst to DeRoche, there were two eyewitnesses: two men having a quiet chat in the shadows. They saw everything as it unfolded. As is often the case, however, these facts, which sound so cold-blooded, become slightly less clear as more details are known about the case.

DeRoche was arrested immediately and placed in the custody of the well-meaning but drunken Officer Farnsworth. Farnsworth, worried for the safety of his prisoner, decided that the officer's own rented room in a boardinghouse would be safer than the jail. He took DeRoche from the jail to his room and shackled him to the bedpost. DeRoche offered Farnsworth $1,000 to let him go. Farnsworth declined, but DeRoche kept him talking until near dawn, when Farnsworth fell asleep. DeRoche then went through Farnsworth's pockets until he found the keys and made his escape, without taking any money or valuables from the snoozing officer.

DeRoche disappeared. The townsfolk were up in arms. The victim, Treloar, was known around town as a fairly harmless man who had been somewhat incapacitated mentally after falling 225 feet down a mine shaft in Virginia City some time back. DeRoche, in contrast, was known as "a petty larceny thief and somewhat dangerous." The popular view was that DeRoche had cold-bloodedly killed a weaker, unthreatening man. A posse was organized.

The town was just as angry at Farnsworth for losing the prisoner. Because tempers were running high, Farnsworth was escorted almost to the state line for his own safety. Meanwhile, the posse searched Bodie high and low for DeRoche.

Because people of the same ethnicity tended to cluster together, the posse hassled another French Canadian in town until he conceded that DeRoche was holed up at his brother's wood ranch outside of town. In the middle of the night, the posse arrived at the adobe cabin 8 miles from Bodie and 2 miles from the Goat Ranch on the Cottonwood Canyon road. When DeRoche appeared, he surrendered, crying, "Hang me! Hang me!" but the posse told him to get his clothes on; they weren't going to hang him (yet).

On the ride into town, DeRoche said Treloar had killed *himself* because DeRoche had grabbed Treloar's arm when Treloar was attempting to shoot DeRoche. When the posse pointed out that he himself had told Mrs. Treloar, "I have shot your husband," he replied he had made a mistake. Indeed he had.

Upon being delivered into town exceedingly early on Sunday morning, 16 January, DeRoche was deposited in the jail, where he was guarded by the constabulary and several citizen volunteers. It was quickly decided that they should hold a court session that day, to better preserve what was left of the suspect's life. Court convened at two o'clock in the afternoon, and several witnesses gave testimony, including the new widow. When DeRoche attempted to retain that savior of defendants, Patrick Reddy, as his lawyer, Reddy declined, stating he had already been hired by the prosecution: a bad turn of events for DeRoche. He selected General John Kittrell for his defense.

Court recessed that evening, to reconvene at nine o'clock the next morning. Unfortunately, several Bodieites ran out of patience and, as the newspapers referred to it, summoned "Judge Lynch" that evening to render a swift and permanent verdict.

There had never before been a deadly serious vigilante group, or "601," as many such groups were called, in Bodie. There had been strident items in the papers calling for vigilante action now and then, but this was the first actuality. The 601 was composed of "leading businessmen" from the town. They took DeRoche to the same spot where he had killed Treloar, and brought over an enormous tripod from a wagon works. The tripod had been outfitted with ropes for two, but Mrs. Treloar's neck was saved by one vote when the 601 polled its members. She was given twenty-four hours to get out of town instead.

After they lynched DeRoche in the wee hours of the morning, they fetched a doctor to verify he was dead, then cut him down and gave his remains to the undertaker. Local citizens took portions of the rope as grisly mementos. When a voice among the 601 lynch mob (generally alleged to have been Patrick Reddy's) offered to pay $100 to anyone who was willing to print his name in the paper as having taken part in this, the mob suggested that *he* should be given the next rope, and so the speaker made a quick escape.

During the brief snippet of trial afforded DeRoche, more details about the fateful evening were forthcoming, but the situation was still less than clear. It seems that DeRoche (perhaps actually Jules Daroche, listed in 1879 in Mono County Records, the *Great Register of Mono County,* as a French Canadian carpenter; no Joseph DeRoche appears)[1] had made the acquaintance of Johanna Treloar, née Lonahan, prior to living in Bodie. They met in Chicago, where DeRoche lived with his wife and three children. Lonahan was in Chicago visiting a married sister, but "became a frequent visitor to DeRoche's house." DeRoche left Chicago for Virginia City, leaving his family behind but intact. Lonahan ventured out to San Francisco. In 1878, Lonahan, with a woman friend, came to Bodie by way of Virginia City. She met Thomas Treloar, a miner originally from Cornwall, England, while she was staying in a Bodie hotel. They were married in January 1879.

The marriage confused some in Bodie, because of Treloar's obvious mental deficiency. Treloar's best man claimed that he had actually asked Lonahan why she married Treloar, and she had replied it was for the insurance money. The record of their marriage license has both Treloar's and Lonahan's signatures. Treloar's is very illegible and scrawled; one would assume from its appearance that the writer had difficulty writing.

DeRoche arrived in Bodie not much later, and, according to Johanna Lonahan Treloar, he was "not . . . in the habit of calling at my house, except in the company of my husband," contrary to implications made in the papers. Mrs. Treloar further testified that the men were sometimes "bad friends" and sometimes "good friends."

It should be noted that Thomas Treloar had been arrested in the past for beating his wife. Johanna Lonahan Treloar testified that she was afraid of her husband when he had been drinking, and she thought he had been drinking that night. She had asked him to go home and start a fire to warm the house, but was afraid to return home after the ball (where she had been washing dishes). Because she was afraid, she asked another man with his wife (T. A. Stephens) to please walk her home. Stephens confirmed this in his testimony. As they passed the corner of Main and Lowe streets,[2] they saw DeRoche being held by police. It was then that DeRoche called out, "Mrs. Treloar, I have killed your husband!"

Although the popular opinion in the press seems to have been that Mrs. Treloar plotted to have her husband murdered for $1,000 in insurance money and to allow her to be with a lover, DeRoche, it may not have been so. Having been beaten before, it seems possible, though far from certain, that she was understandably frightened by her angry, potentially drunk husband that evening.

It would appear that DeRoche was guilty as sin. As Treloar left the ball, DeRoche slipped out after him. However, he, too, has a case slightly murky with complications. Two lawmen testified that Treloar had complained to them that DeRoche was interfering in his marriage. One of them said Treloar had threatened to kill DeRoche, something not taken lightly in Bodie; however, neither was interfering in a marriage. "I told DeRoche Tommy meant business," said Captain Morgan. It is interesting that in this case, the usual Bodie attitude toward threats appears to have been suspended.

One of the witnesses to the shooting said DeRoche and Treloar were quietly walking side by side when Treloar stepped down off the sidewalk, "thus getting a little ahead of the other (DeRoche), when the other man fired. . . . They were not quarreling when they passed me."

The unwillingness on the part of the 601 members to wait for due process of law, while unfortunate, is probably due to several factors. A small, seemingly harmless man (though not so harmless to his wife) had been murdered by a larger, more capable man (whom Treloar had threatened with death in front of witnesses). Furthermore, it involved a wife and money and, worst of all, took place in the "proper" section of town. Bodie juries were famous for acquitting defendants, so there was also good cause to expect DeRoche would go free. The enormous pressure of public outrage must have played a part as well. At Treloar's funeral, the minister, the Reverend Warrington, said, "[I]f a man have an irresistible impulse to take another man's life, I say let the law have an irresistible impulse to put a rope about his neck (describing the operation with his finger) and take the life from his body" (McGrath 1984, p. 238). When the minister is calling for your death penalty, you know you're in trouble.

As for the widow Treloar, an item in the town papers several weeks later noted that she had been seen in Virginia City, but there her trail grows cold. And so ended the sad and, fortunately, solitary story of Bodie vigilante justice. Bodie now had its church and its hanging, which many '49ers would tell you meant the town was headed for decline.

SPRINGTIME

In February 1881, as the winter snows melted, the ongoing problem of flooded streets appeared again, making the roads "virtually impassable . . . on Thursday the river running through Mono Street would almost swim a horse, in places" (*Daily Free Press* 4 February 1881). One newspaper item (*Bodie Morning News* 14 May 1880) claimed a mule had disappeared up to his ears in mud! Solutions were debated, and when sewers were deemed too expensive, the citizens pondered digging and maintaining an open ditch from one end of town to the other.

In March, Jailer Kirgan, a fixture on the personnel roster of Bodie law enforcement, and concierge at the "Hotel de Kirgan" (jail), died from a horse carriage mishap. He was thrown from his carriage when the horse bolted, and landed hard on his head on a frozen street. He lingered a few days, then died. The town was very disappointed to lose such a steady character.

The same month, as the mining bonanzas seemed to have come to an end, various companies began preparations for "deep mining" with the customary "bright outlook" predicted for the spring in the *Daily Free Press*. The growing concern about Bodie's future wealth was not relieved this spring. The San Francisco Stock and Exchange Board's continued lackluster performance depressed Bodie stock prices (and investors), which resulted in the halting of work in nine mines in April 1881, including Bechtel, Belvidere, Summit, Double Standard, Dudley, Champion, Goodshaw, South Bodie, and University.

Just in time, the Cook brothers, of Standard Company fame, decided to finance (with other capitalists) the Bodie Railway and Lumber Company railroad project, from Bodie Bluff to Mono Mills, the sawmill on Mono Lake's southeastern shore (*Daily Free Press* 30 April 1881). The Bodie terminus was to be at the top of Green Street. (One of the railroad's old buildings still stands there. Although State of California maps label it the "Railroad Station," some say it is actually the old office building.) The Cooks maintained that their $500,000 investment in the railroad demonstrated their faith in the longevity of Bodie. Bodieites looked to the project with hope and enthusiasm.

Rail workers were offered $1.25 a day plus board to work on the project. Many of the Bodie unemployed balked at the rate of pay, knowing full well that miners could bring home $4 a day. Still, there was not enough mining work to go around, so several men did sign up for the job (Wedertz 1969, p. 159; *Daily Free Press* 25, 28 May 1881).

In May, about 68 Chinese and 260 white men came to work on the project. When Anglo Bodie miners noticed the group of Chinese men hiking toward the railroad work site, they became enraged. They hastily called a meeting and resolved to banish the Chinese workers from the project. When they demanded that the superintendent fire all Chinese workers from the job, he claimed he needed a few days to decide. The threat of trouble was great, and in town the newspaper reported that should any problems arise from this, "every mine in the district . . . would be closed for about 90 days." The paper further noted that because the Standard Mine was the only dividend-paying mine, it would be the one with the greatest immediate loss. However, because the Standard's owners, the Cook brothers, were also the financiers of the railroad, they would not be likely to bow to the pressure of a mining strike, and, "In the meantime, the whole community would get uncomfortably cool." The railroad refused to sack the Chinese workers. Instead, they pledged to hire "all the white laborers we can get." A group of about 40 "excited individuals" (who were white) headed angrily for the work site (*Daily Free Press* 25, 27 May 1881).

Luckily for the workers themselves, word preceded them, and the men overseeing the railroad project hastily gathered up the Chinese workforce and took them to Paoha Island in Mono Lake. (There are two islands in Mono Lake, one very light colored, one very dark colored. Paoha Island is the lighter of the two.) There they camped out with thirty days of rations and waited.

The journey from Bodie to Mono Mills was neither easy nor short. As the trek wore on, the angry Bodieites wore down and the purpose of their trip seemed less clear and pressing. Many changed their minds about the excursion and returned to town. The smaller ragtag group that finally reached Mono Lake found no Chinese workers to assault, only a semiabandoned work site. The *Daily Free Press* (27 May 1881) noted, "[T]here was nothing left . . . but to stand on the shore and make faces at the . . . Celestials who were cooking gulls' eggs in the hot springs on the distant island."

After a show of empty threats from the Bodieites, which were met by solid threats of arrest from the sheriff, the group of "largely ignorant and weak-minded laboring men . . . egged on by a few bankrupt woodmen," as the newspapers described them, returned to Bodie. The Chinese workers returned to work, and life went on as usual with no further demonstrations about the topic.

GETTING BY

No new bonanzas of salvation appeared in Bodie. As onetime excitement faded into just getting by, newspapers hinted at the gradual changes in Bodie's atmosphere. There were many notes about Bodieites who had moved on, describing where they'd gone and what they were doing in their new locale. Many of these articles focused on a particular location and who was there, such as "Bodieites in San Francisco." The newspapers also periodically reiterated how much better off Bodie was with fewer riffraff in town. As some of the rougher element departed Bodie for other parts, the number of Bad Men from Bodie appearing in other towns increased sharply. The Bodie papers tried to keep up with the antics of these braggarts, who generally favored storming into saloons and bellowing about, for show.

In May, a traveling psychic came to Bodie, holding appointments at a local hotel. One can only imagine how her work was cut out for her in the declining mining camp. It is not noted whether she foretold of the ten additional mines that would stop work that same month. These closures inexplicably prompted papers to opine that it was clear that livelier times were on their way, and the people turned their hopes to the Bodie railroad construction as an indication that they would pull out of this slump.

When the 1881 Fourth of July rolled around, people were cautious in their festivities because President Garfield had been shot and was in dubious condition. When Garfield rallied, they went ahead with the celebration. It was a bit smaller than the one in 1880, with only one brass band. But the other traditional festivities went on, including wrestling matches, literary events, the ever-popular horse racing, baseball, and so forth. Bodie's Fourth was still a grand celebration. (Unfortunately, President Garfield died later that year.)

Ironically, despite the closure of mines, the lack of bonanzas, and the trickling outflow of residents, the Bodie Mining District's actual gold production was actually *up* at this time. In early August 1881, the *Daily Free Press* reported, "The Bullion shipments from the district . . . the first six months of 1881, were $1,600,000. The shipments for the same period in 1880 were $1,400,000 and in 1879 they were $1,100,000. . . . [N]otwithstanding the thinning out of population, and the low price of stocks, Bodie is now producing a better average of bullion than ever before." But still the people left Bodie.

By the hot, dry August of 1881, most of the remaining mines were up against an unlikely obstacle: water. The water table began to show itself at the 450-foot level in virtually all Bodie District mine shafts. Bodie's irony lay in the fact that the town could be so arid on the surface, and yet an abundance of water *under* the surface would be a major factor in the abandonment of its riches. It was very expensive to pump water out of the shafts to get to the ore. This quickly separated the big players from the smaller ventures. Only the richest mines could afford to pump out their shafts. Those who had just been eking out a small profit had to sell out or give up once they reached the watermark. By midyear only a few mines were still clearly worthy of continued effort. However, those few mines required many employees. Those employees, plus the supporting businesses and a few more determined prospectors and miners, made up the town of Bodie after its fall from boom-time grace.

Bodie's "decline" was so gradual, the term may imply more than it actually meant. During the boom time, Bodie had swollen to an unreasonable size for the work available there. In 1881, Bodie was still a vigorous town with a fairly healthy economy, provided the population did not get too large.

If one were looking for an omen in Bodie, the autumn of 1881 would appear to have provided a few. In mid-October, the sheriff invoked a new California law and ordered all the opium dens in Chinatown to shut down. The fine for smoking opium was $500, and for selling, $1,000. This removed a major source of income for many Chinese Bodieites, inspiring some of them to head for greener pastures elsewhere.

The same month, in a classic display of timing, a temperance lecturer came to Bodie and held forth about the evils of alcohol. Most of the local drink masters attended the lecture, and, despite the lecturer's fiery warnings, they were nonchalant. One fellow was unconcerned about the effect of consciousness raisers on his business, maintaining he could mix more drinks than they could deliver lectures. Another barkeep said he had no problem with the lecturer, except when he exposed the secrets of whiskey making: "How simple and how cheap and the cost. I really thought it was unfair . . . to give that away."

As if the closing of opium dens and the appearance of a temperance lecturer weren't enough for the Bodie "bad men," a California law was enacted that forced saloons and cigar stores to close on Sundays. It would be an understatement to say the miners disliked the Sunday closure law. The local district attorney did too, and refused to prosecute people under the law. He lost his job as a result. Mr. A. I. Weiler kept his cigar store open and was hauled in front of the justice of the peace. When he demanded a jury trial, "after exhausting two venues, only six men could be found competent to try the case."

A RAILROAD AT LAST

On Monday, 14 November 1881, at 3:00 P.M., the last spike was driven into the Bodie Railway and Lumber Company's Bodie terminus, and two locomotives chugged into the station. A *Daily Free Press* item reported that the Kuzedika "have become reconciled to the new railroad and look forward with pleasure to the time when they can ride from Bodie to the lake." The same year, the railway to Mono Mills was completed, enabling the mines to receive steady shipments of the all-important wood at a lower cost.

Wood brought in from Mono Mills went not just to Bodie, but also to the neighboring mining camps of Aurora, Masonic, and Lucky Boy, transported there from Bodie using the traditional mule trains. Not long after the railroad began its lumber deliveries, the *Rocket* was retired from its

ABOVE: *The railroad meant timely and plentiful wood supply at long last. Courtesy the Bell Collection*

BELOW: *An unidentified man standing at the Bodie terminus for the railroad, up on Bodie Bluff. Courtesy the Bell Collection*

sometimes-dangerous Mono Lake runs. The completed railroad cheered the townspeople. Soon there was talk of establishing a railroad connection to the outside world via the Carson and Colorado Railroad at Benton. The name of the railroad was eventually changed to the Bodie and Benton Railway and Commercial Company. People felt this would maintain Bodie's stature as an established settlement. Or at least ensure her survival. Many saw this plan as Bodie's salvation.

HOLIDAY CONSUMPTION, BODIE STYLE

Thanksgiving of 1881 was an elaborate holiday. Although the town population was lower than the previous year, friends and family put away an astounding amount of food. Some people ate as many as thirteen meals "during the day and evening" according to the *Daily Free Press,* which reported that altogether, Bodieites and guests consumed "31 beeves, 182 sheep, 212 hogs, 1,386 turkeys, 1,121 chickens, 3,824 lbs. of fish."

The Bodie Mining District was still among the top producing districts on the Pacific Coast. During the course of 1881, over $3 million in gold bullion was shipped out of Bodie. And yet, she continued to decline. In December, the trickle of people leaving Bodie changed in volume and kind. More people left, and not all of them were fair-weather citizens. Bodie's shifting economy—mines closing down and no new bonanzas—made points elsewhere more appealing to many. The completion of the Bodie Railroad also meant the end of many jobs. Chinatown fell on harder times after the closing of the once-thriving opium dens. On top of all this, the stock market by the Bay continued to flounder. Once it seemed clear that Bodie had no alternative focus outside of mining, it appeared prudent to many to consider moving on, especially before winter set in.

1882: THE WINTER OF BODIE'S DISCONTENT

The rapid increase of dogs in a mining camp is said to be a sign of business decay. However this may be, Main Street is the playground for many a brute.
—Daily Free Press *(January 1882)*

A hard winter started 1882, and it seemed to bode poorly for the town's future. As time dragged on with no new strikes, it became increasingly apparent that, out of the almost sixty Bodie District mines, only the Standard and the Bodie would remain steadily profitable for any time. Bodieites also became more aware that the town penchant for feverish investment in the newest strike or rumor was fattening up businessmen on the Barbary

Coast but not helping Bodie or its citizens. The jubilation of the boom years faded as more people hit the road.

People in debt frequently fled Bodie in the middle of the night, heading for the state line. The stage that left Bodie at two o'clock in the morning was very convenient for these people. The *Daily Free Press* even went so far as to suggest building a small "custom house" at the state line to try to halt those fleeing to avoid paying their debts.

In January the bullion shipment figures for 1881 were released to a shrinking readership: "It will perhaps be a surprise . . . but . . . bullion shipments from Bodie during the year just closed were the largest of any year since the revival of the district. They amounted to . . . $3,173,000 . . . more than $100,000 over 1880, . . . more than $600,000 over 1879, . . . more than a million over those of 1878, which was the year of greatest excitement in Bodie mines." The papers all expressed great optimism about the future with the usual exciting predictions. Just two days later, the same paper reported, "There is not much demand for lumber just at present. Building is at a standstill. Those who want a house either rent or purchase one already waiting for an occupant."

As the population dwindled and the "bad men" fled town, Bodie settled down into a more placid small town with fewer excitements and less chance of random danger. The local justices felt the difference; their workload was significantly reduced. The lawmen could also relax a little. Between the reduced population and the increased wood transport, thanks to the Bodie

Bodie Gentilities: Whipping and Drink Making

Even though Bodie had become a smaller town, the general attitudes about proper behavior persisted. An angry mother, accompanied by a doctor, came into the Wells Fargo office one February afternoon in 1882. She approached a twelve-year-old clerk and told him if he ever spoke "in vile terms" to her daughter again, "she would reduce him to a corpse." She then lashed him with a small whip, cutting him. As the story went, the boy had been rude to the woman's daughter, and when he was asked to stop, only got worse. He, of course, said it wasn't him, it was the other boys. The *Daily Free Press* immediately called for the end of a gang of "hoodlums" who hung out on Green Street and hassled girls. Perhaps that would have been the end of it in most places: not in Bodie. The whipped boy's mother, incensed at her son's punishment, stormed downtown, bought a new whip, strode into the office of the doctor who had accompanied the girl's mother, and proceeded to whip *him*. The doctor took the whip away from her and then put it on display in his office.

Just a few days later, "The Bodie Society for the Promotion of Genteel Habits and a Higher Civilization" met. They did not discuss the proper method of polite flogging. Instead, they pondered, "What should be done with the piece of lime found in a cocktail." After due consideration, they sided with someone claiming to know the eastern and European way: "[T]he lime should be fished out, dropped on the tongue in as graceful a manner as possible, and after being firmly pressed . . . placed back in the glass" (*Daily Free Press* 10 February 1882). Everyone agreed this was best, because the barkeeper could use it again for the next cocktail.

Railway and Lumber Company, there was more than enough firewood on hand for everyone, and the annual panic about it was laid to rest for now.

The railroad closed down for the winter in early January, finishing the season with a special excursion for railroad superintendent Thomas Holt and some of his friends. After outfitting a flatcar with an enclosure, benches, and stove, they headed out to Mono Mills for the day. Once there, they amused one and all by talking with people back in Bodie over the telephone, a fascinating novelty at that time. Those in Bodie sang a song for those in Mono Mills; then the people at the Mills reciprocated. After this excursion, both the railroad and the mills lay idle until spring.

In early March, anti-Chinese feeling surfaced again, especially among the many unemployed white men, despite the fact that many Chinese had left Bodie. The governor of California fueled the fire by declaring 4 March a legal holiday "for the purpose of allowing the whole people . . . to discuss the Chinese question." They did so in Bodie, and drafted a resolution to send to Washington, D.C., urging an end to admitting Chinese to the United States. Out of 1,200 votes cast in town, only two were pro-Chinese.

BODIE'S DREAMS FADE

The Independence Day of 1882 celebration was much smaller than in years past, but with the usual athletic events and merriment. The week after, the railroad work to extend the Bodie and Benton Railway from Mono Mills to Benton was halted, dealing a strong blow to Bodie morale and dreams. No official explanation was forthcoming, but some supposed that those who owned Mono Mills did not want the rail connection to be completed because it would undercut their other lumber interests in the Lake Tahoe area.

The Mysterious Dr. Blackwood Redux

The late-winter doldrums were abruptly interrupted when a grisly mystery surfaced in Bodie. A reporter was taken to a house in town and shown dismembered human remains that had been dissected and experimented upon. Shortly thereafter, more remains were found frozen, dumped in the Ajax mine shaft. A general frenzy ensued. No citizens appeared missing. A small group examined two recent graves and found them undisturbed, but a recently bereaved widower became concerned and went with a group of friends to check his wife's grave. They found it disturbed and his wife's body gone.

It was determined that Dr. Blackwood, the same doctor who in 1880 had fenced off Standard Avenue in the middle of the night, had done the grisly deed. Although the widower's wife had died naturally, Dr. Blackwood had tried to further his scientific research endeavors after the fact. While the grand jury debated what to do, the doctor fled town, never to be seen again. Bodieites were both horrified and stymied by the whole affair. Thus ended the Bodie saga of the curious Dr. Blackwood (*Daily Free Press* 19–26 February 1882).

Real estate prices plummeted. Houses were selling for 25 percent of their original price. People continued to leave in numbers, and the overall spirit of Bodie was downcast. Near the end of the year, the mining complex that included the Noonday, North Noonday, and Red Cloud mines was attached by Wells Fargo for back payments, closing down their operations and putting 175 men out of work. The sheriff shut down the Bullion Mine. As soon as the pumps were stopped, these mines, like others before them, quickly filled with water from the water table.

In September, local Catholics opened the doors to their new church. A week later, the Methodists followed suit. It is hard to discern if this was a case of "a day late and a dollar short," or if it indicates that the original spirit of Bodie had passed away at last. People tried to keep a cheerful countenance, but at times it was difficult:

About the most solemn and dreary spot in town is the Chinese quarter on King Street. The members of the two great companies have ceased to fight, the gong is muffled, and by midnight only the white opium fiend can be seen coming and going through the narrow street. Where $2 was spent for the fatal drug three years ago, not two bits are squandered now. The questionable glory of Chinatown has departed; its brightness has been obfuscated, and the buildings with all their crooks and crannies are covered with fuliginous[3] matter. (*Daily Free Press* 15 December 1882)

As more mines shut down and more miners were paid in IOUs instead of hard cash, the mood grew tense in Bodie. The same week that it was announced that the sheriff would be selling the Oro, Noonday, North Noonday, and Red Cloud properties, a miner named Mike McCallum took his payday matters into his own hands. An employee of the Oro Mining Company, he was told that he, like the other employees, would be receiving one-third of his amount due, with the balance to be paid sixty days later. McCallum, mindful of the times and the likelihood of ever seeing his full pay, pulled out a foot-long six-shooter and told the superintendent that he wanted *all* his due pay. He left with a check for the full amount and promptly cashed it. He was later arrested for robbery, but was acquitted by Judge Phlegar, to the great approval of onlookers in the courtroom.

Tax assessments rose and were soon larger than the dividends. When taxes due were exceeding dividends earned for investors, more mines shut down. Over the course of 1882, several mines disappeared from the San Francisco Stock and Exchange Board, including Red Cloud, Maybelle, Concordia, Belvidere, Blackhawk, Noonday, Booker Consolidated, Pacific, Dudley, North Noonday, Boston Consolidated, Oro, Paris, and the Addenda. There were

500 miners employed in the northern mines and 175 unemployed miners in the district. In 1882, the year's total gold production in the Bodie Mining District was slightly more than $2 million. By the next year, it had sunk to less than $1 million.

As of 31 December 1882, the total amount paid out of the Bodie Mining District from its beginning was $13,937,832.04 in 1882 dollars and cents[4] (Wedertz [1969] 1986, p. 152).

1883–1894: DOWNSIZING

To a Bodieite a visit now to Bridgeport is like escaping from a prison.
—Bridgeport Chronicle-Union *(1 March 1884)*

As of March 1883, the mines were producing $150,000 each month in bullion, but only three were paying dividends: the Syndicate, the Bodie, and the Standard. The bullion shipments were smaller than in years past. Stock speculation was way down, and the economic structure of the mines and companies meandered toward further consolidation.

Between 1883 and 1895 Bodie continued to shrink, but it did not really threaten to disappear entirely. There were still enough people living in Bodie to constitute a real town, with businesses and a schoolhouse and the ever-present mines and mills. As 1883 began, the whole town numbered about 3,000, with about 500 men employed in the mines and mills.

After increasing incidence of mischief, some malicious, including the previously mentioned incident of boys attacking a Chinese boy one night as he walked home, the town put a curfew into effect in early 1883, requiring all boys to be home by 9:00 P.M. unless they were out with their parents. This was in stark contrast to the "anything goes" spirit that pervaded Bodie life in its boom time.

By midwinter 1884, only six or seven mines in the entire district were still in operation, and Standard Consolidated stopped paying its famous dividends. Bodie's population had shrunk to 1,500.[5] In 1884, five years after the boom, the vacant buildings in town outnumbered the inhabitants three to one:

Many of the buildings are going to ruin; the sidewalks are cranky; unlatched doors swing on rusty hinges; demolished stovepipes sway to and fro and here and there the legends "for rent" and "for sale" stand out with great prominence. Nothing is quite so depressing as a row of deserted houses—unless it is an empty stomach. (*Daily Free Press* 28 February 1883)

The growing number of vacant buildings in Bodie presented another financial opportunity for those with initiative: Many men began tearing them up and carting the lumber to Hawthorne, Nevada, for sale. In light of this new enterprise, the Mono County district attorney was moved to notify people that they could not remove "improvements" on real properties that had delinquent taxes because the lien on their property included those improvements. Violators were threatened with a year in jail, a $5,000 fine, or both. Therefore, people who had property in town but could not afford to pay the taxes could not tear down their own property and sell the materials to pay their taxes, either. Some people simply abandoned their property altogether. Delinquent Mono County tax lists frequently filled the *Daily Free Press.* J. S. Cain purchased several of the houses for the price of their back taxes.

Bodie managed to hang on as a small town, with small-town joys and triumphs, laying aside the big-city dreams she once held dear. The life there was comfortable enough, though with the usual struggle against the elements. Still, Bodieites dreamed and searched for a way to bring their mines back to life and begin another bonanza.

THE GAY '90S: BODIE IN FLUX

The 1890s brought technological innovations everywhere, including Bodie. Although the town's heyday was over, some of these new technologies brought some excitement and hope to the mining town.

Hydroelectric Power

In 1892, Bodie saw the construction of the world's first hydroelectric power plant and transmission line. This was another venture supported by J. S. Cain of the Bodie and Benton Railway and Commercial Company fame (and owner of increasing numbers of Bodie properties).

Mining engineer Tom Leggett was the design engineer for the project. He had been casting about for lower-cost power solutions for the mines and mills, and noticed that in Telluride, Colorado, a power plant with a three-mile transmission line had been put in. He consulted with General Electric, but they were "still wedded to direct current." He then talked with Westinghouse. Together they did it, "using a 250 kilowatt generator at the water power end, direct connected with Pelton water wheels under 300 foot head, without transformers, the current being generated at 3,000 volts and carried on No. 1 bare copper wire to Bodie, where it was applied to the operation of the mill," said Leggett (Rickard 1922).

The power generated at Green Creek was transmitted to Bodie via 13 miles

Hauling in hydroelectric parts in the 1920s. Courtesy the Dolan-Voss Collection

of perfectly straight power lines, no curves, because the popular notion of the day was that electricity couldn't travel around curves. The experiment was successful, and the Bodie mines were fairly quickly relieved of their reliance on wood. Electric motors replaced steam power in milling and mining operations. The Standard Mill's steam engine was sold when the firm converted to electricity: There was no turning back. Although the mines had electric power, most town residences didn't have power until the 1911 construction of the Jordan power plant near Lundy. Bodie maintained its priorities, even after the boom flush had died down.

The Green Creek plant was heralded the world over, and its designers were in great demand, eventually being hired by the British government for similar work in British colonies such as South Africa.

The Cyanide Process

The 1895 invention of the cyanide process revitalized Bodie mining to a small extent. Tom Leggett, design engineer for the Green Creek power plant, was interested in the potential of the process to make Bodie's enormous tailings piles valuable. He sent a sample of a ton or two to the American representatives for MacArthur-Forrest in Glasgow, Scotland, where the cyanide process was originally developed. Their results were poor, so they left the idea alone.

The next year, Leggett met Alexis Janin in San Francisco, where Janin was working on the process with Charles Merrill of the University of California. He sent them a tailings sample, and they reported much better results. Leggett then asked Merrill to come to Bodie, where they tested 8 tons in all, from various areas. Their results "were so good that we erected the first 100-ton capacity cyanide plant on the Pacific Coast, work being started in June 1894, the plant completed in September, and paid for out of the proceeds by early December, when it was forced to close down on account of the severity of the weather," Leggett recalled in a 1922 interview (Rickard 1922).

Another account (Billeb [1968] 1986) claims that a Virginia City fellow, A. J. McCone, came to Bodie and proposed to J. S. Cain that they construct a plant to use the new process. He expected they could remove between 90 and 95 percent of the gold from the ore. J. S. Cain agreed and promptly purchased the "worthless" Noonday tailings piles while McCone paid for a cyanide-process expert to help them construct their plant. The cyanide plant and process were supposedly very mysterious and secret. However, at least one townsman peeped through the fence and learned enough to figure out the rest and built his own cyanide plant at the Syndicate Mine. Later, the Standard Company built a cyanide plant as well.

Although the cyanide process did not bring about another boomtown, it did revive some interest in Bodie. The tailings piles from earlier mining were all gone through using the cyanide process, wringing the last of the gold out of the rock.

7

"Up home" meant going to Bodie to the old timers.
—*GORDON BELL, former Bodieite*

Up Home

The twentieth century brought significant change to Bodie as everywhere else. No longer a thriving boomtown of thousands with more arriving
each day, Bodie settled into being a small town with a few steadily working
mines. The essence of Bodie, however—the rugged climate, the hardworking
and hard-drinking men who mined there, the possibility of danger in any
given moment—remained, even as the number of residents decreased.

Several people who were born in Bodie, and others who spent time there
growing up, were pleased to have an interviewer give them a chance to take
a look back to their times up home in this town. The interviews that make
up this chapter come from personal interviews I conducted and from Bodie
State Historic Park archives. (The park archives include personal letters written as well as some taped interviews. All the interviews can be found listed
in the references section of this book.) These reminiscences are from people of varying ages, and describe different times, from the early 1900s clear
through to the turn of the twenty-first century. But all these people remember Bodie vividly as *home,* not a ghostly remnant of the Wild West.

When you talk with these former Bodie citizens, you hear a consensus
that Bodie was a wonderful place to grow up. People felt safe and part of a
community. Bodie was still a town where no one locked their doors and children roamed the hills at will, their parents free of worry about kidnappers
and other atrocities.

Pat Goodwin commented, "One summer my cousin returned home to

Oakland after some time in Bodie and shortly thereafter complained to her mother, '[T]here's nothing to do here.' Her mother replied, '[H]ow can you say there's nothing to do here? You just returned from *Bodie!*' She said, '[Y]es, but there, it's a different *kind* of nothing to do!' "

Katie Conway Bell Adair's parents lived in Bodie at the turn of the century; her father had first arrived in the late 1880s. She recalled, "My folks lived . . . in Green Street there . . . close to all the bars, and they'd hear a shot go off and my father'd say, 'I better go see who's gone, 'cause by morning they'd get fighting in an argument and they'd shoot somebody. They'd come off the shift from the different mines and they'd all argue and fight."

Katie was born in Bodie, but not with the help of an attending doctor. "My father's partner's wife dropped me into the world, . . . Mrs. Dixon. . . . The thing that really got the children in Bodie was typhoid fever." Katie Conway is the only Bodieite to mention typhoid.

THE HORSELESS CARRIAGE ARRIVES

On 12 June 1905, the town greeted its first automobile, courtesy of Bodie resident Cecil Burkham. Because cars were so new, and traveled on the rough and bumpy stagecoach roads, most automobile trips required a substantial supply of repair materials. Whereas wagon wheels ran over glass and sharp objects with little difficulty, automobile tires were punctured. Few

The horseless carriage arrives in Bodie. Courtesy the Dolan-Voss Collection

The prices on this Bodie gas sign verify the sign's antiquity!

people left on car trips without a supply large enough to render them a rolling garage. There were no mechanics. It was up to the owner to fix the car on the spot, or walk. Naturally, people regarded these mechanical contraptions differently from the way they do today.

Emil Billeb, superintendent of the Bodie and Benton Railway and Commercial Company, illustrates this situation in *Mining Camp Days* ([1968] 1986) with the winter tale of how he and a friend had to leave the friend's broken-down car and walk home in the snow. When they returned to the car, some three weeks later, they could not get it running. In disgust, they built a fire under the car, intending to burn it. They stood back for the show, but instead the fire burned itself out. When they returned to the car, it started right up, so they drove it back to town.

Although cars gradually took over in Bodie as everywhere else in the world, Bodie's terrain and weather slowed the transformation there, and inspired some creative variations. The town's mechanical whizzes fashioned several mutations of the automobile to try to accommodate their needs. There were cars outfitted to run on railroad tracks, and various attempts were made to turn cars into Bodie snowmobiles for the wintertime. The snow cars showed all kinds of ingenuity, from front wheels replaced with sleigh runners (somewhat successful) to triangular wheels that would walk over the snow (not as successful). When people nowadays wince over the last three miles of Bodie's dirt road, they would do well to recall the Bodie citizens traveling in early cars over nothing but dirt stagecoach roads.

SMALL-TOWN LIFE

Howard Ball's father was the newly ordained Methodist minister assigned to Bodie for the year of 1905. Now, cars may have appeared in Bodie, but they were not yet commonplace, and the Ball family arrived on a stagecoach. Howard remembered the long ride into Bodie:

We went by stagecoach two days and one night to get to Bodie. . . . [T]he way that they had to get [there] from Reno was through Carson City, Gardnerville, and then in through the passes which are now highway 395. They left the main road to climb the hill to Bodie. . . . Every two miles there was a stage stop where they put on new horses to climb the steep grade to Bodie. . . . It got so cold, the stage driver went into one of the stops and got a buffalo robe and wrapped it around me and so I sat between my mother and father and finished the trip up to Bodie wrapped in a buffalo robe.

Winter transit remained horse-based for many more years. It was hard enough with horses, impossible with cars. Courtesy the Dolan-Voss Collection

The winter that the Ball family spent in Bodie was a difficult one. Howard recalled a wood shortage that quickly grew serious:

Dad asked the people from the whole neighborhood to bring a stick of wood to the church so that we could all use the same heat, to conserve fuel around the town. . . . The big snowstorm came at that time . . . before the men could go down the mountain to get wood. . . . It was . . . necessary for us to stay in the church for eight days and nights. There were a good many children and we had a great time sleeping in the church . . . you might say we were almost camping in the church . . . 'cause nobody went home because the houses were all cold. I remember that my place was sleeping on the left side of the church in the second pew from the front. . . .

Another interesting thing . . . in wintertime when they couldn't get meat, why if a man came in with a deer or a bear, they rang a bell and they opened up a little building and everybody came there to get a share of the meat, which was needed so badly . . . and Dad being a minister, of course, he got mostly the hearts and the livers.

The ice from that winter lingered for months. The following summer, Howard Ball's father made use of it: "[T]here [was] a pile of ice remaining from the winter. He chopped it out to make ice cream for an ice cream social over in the church on July 4th for their celebration."

Katie Conway recalled winter precautions: "When the men were working in the mines and it snowed, the folks at home would put a colored calico on the roof so the miners would know where home was when they got off work."

Jack Bryson was born in Bodie in 1901 and lived there until he was ten. He recalled winter food preparations included filling their lean-to with "nine sacks of potatoes, and sacks of carrots and turnips. And by getting meat and potatoes, that's how we lived." They also kept a cow in a barn just southeast of their house.

In Bodie, winter was treacherous for people trying to work and provide for a family, but for the children it was great fun. Mrs. Lauretta Gray recalled that in her childhood winters, they'd toboggan down the hills behind the cemetery. "Oh, sometimes we'd be 14 kids on a single toboggan!" She explained that because there were no trees, there was no serious danger. "It'd take about two minutes to go down, and about an hour to climb back up again!

"We used to go skating at Warren Loose's place. . . . Emil Billeb's brother used to come along and pick us kids all up and we'd go down. . . . They had this reservoir. . . . They would cut [the ice] for the saloons, for the ice for the drinks, but before they cut it, they let us go skating."

Howard T. Ball

When Howard T. Ball visited Bodie SHP in October 1977, he described a very different Bodie to Ranger Ken Featherstone. Howard Ball's father, Francis E. Ball, spent 1905 as the minister in Bodie's Methodist church. The church building still greets visitors as they enter the townsite.

As Howard Ball remembered it, a minister's life in Bodie was not lucrative, but it was interesting. His father "had no salary when he took the church at Bodie. His only pay was a $20 goldpiece for each wedding or each funeral. He said at the time this was one of the best charges that he had because he buried so many people up in Boot Hill. . . . They would ring a bell when there was a funeral and they would toll the bell the number of years the party had lived and so the bells were tolling a great deal of the time. . . . [Once] when they had a gunfight on the street here at Bodie and I peeked around the corner—the kids are not supposed to be around but I did, I peeked around the corner—and saw them shoot at each other and one of them was killed.

"It reminds me . . . I remember the funeral of a lady. We could hear the moaning and the crying and everything two or three blocks away. I remember that Mother went over to see the family and talk to them and thought she would try to comfort them and we found out that the noise and all was from some hired mourners. There were two or three women who were hired to mourn at the time of the wake and they mourned for hours and hours at a time, and that was supposed to show the respect for the dead. . . .

" . . . [A]nother interesting thing about Bodie that Dad told us was that the church was built by the money raised from the saloons and the houses of prostitution and the gambling dens and the opium dens. There was a large Chinese population here and opium was widely used. Bodie was considered one of the sin towns of the world at that time."

Independence Day remained a big holiday in Bodie. Jack Bryson remembered the early 1900s: "[O]n the Fourth of July . . . my sister and I were given two dollars. It took us the 4th and 5th to spend the two dollars at all the stores around here. We'd buy sarsaparilla and firecrackers."

Mrs. Lauretta Gray smiled at the mention of the Fourth. "We had a big float and us kids rode on the float. We'd go from Bodie to Bridgeport and all around. Dressed elegantly, best we could in those days. Then we had a baseball game . . . we had footraces. And I skinned them all! . . . They had [miners' contests], and they had potato races and sack races . . . and of course, fireworks."

Jack Bryson's recollections of his Bodie childhood paint a picture of life before modern conveniences. "Cook on a wood stove, take your bath Saturday night. Wash stuff by rain. I always had second-class water. My sister would take a bath first, and then I'd take a bath in her water . . . everything was washed on a washboard." Of course, there were special occasions as well. Jack's father was the caller at the famed Saturday night dances, and at the Christmas parties for Bodie children, Jack said, there were two Christmas trees (not a small feat for a town as treeless as Bodie), and Santa always arrived on a sled, with a stocking full of candy for every child.

Although the most infamous troubles that occurred in the mines were during the boom time, there were also some skirmishes here and there later on. Allen Nutter was born in Bodie in 1906, the son of Edward Nutter, superintendent of the Standard Mill (working with Theodore Hoover, general manager of the mine and President Herbert Hoover's brother). By 1906, the mining properties were further consolidated. The Standard Mining Company included the Standard, Bodie, Bulwer, Tioga, Bechtel, Bodie Tunnel, and Syndicate. These mines, along with the Noonday and Red Cloud properties, were the only claims still operating.

A footrace on the Fourth of July, in the early 1900s. The girl out in front is most likely Lauretta Miller, later Mrs. Lauretta Gray. Courtesy of California State Parks, 2002

About that time the mine was having both labor and high grading problems.[1] A few threats were made against the management, and some dynamite was actually exploded under the mine superintendent's bedroom, blowing the bed and him to the ceiling and back, without injuring him. So my father then had steel plates fixed around the four sides and bottom of my crib, just in case. He also retained a well-known gunman to guard our place. He occupied the same room with me. Anyone entering the house found themselves looking down the barrel of his .44, including Mother.

The only occasion he had to use that gun was when someone shot the local butcher dead in his shop one morning, for dancing too often with his wife the evening before at a ball in the Miners' Union Hall. He was seen going up the hill and entering one of the mine portals. Dad told this gun toter to "go get him," so he did. He entered the tunnel with drawn gun and heard a sound in the dark, off to one side, at which he instantly fired. They brought the butcher killer out feet first.

When Mrs. Gray was growing up, in the early 1900s, the saloons were no big deal: "There were several in town, but I never went inside any of them." In general, women didn't go into them, just as in the boom days. Her father didn't frequent them either. As for opium, she says it was still there in town, but pretty much "just used by the Chinese."

Her daughter, Fern Gray Tracy, comments, "Mom's sister . . . was married to Mike, who ran one of the saloons in Bodie. He named himself and his res-

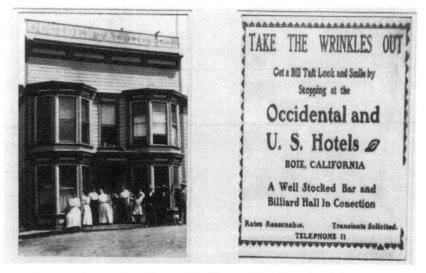

Mrs. Gray's mother, Annie Miller, outside of her Occidental Hotel with her staff. On the right is an ad for the hotel. Courtesy the Gray-Tracy Collection

taurant 'Bodie Mike.' From there he went to Lee Vining and had a bar. . . . I think it's still called that."

Mrs. Gray's mother, Annie Miller, owned the Occidental Hotel and, later, the boardinghouse on Green Street that still stands. When Mrs. Gray was a young girl, she helped out serving in the hotel. "I worked for my mother all the time, until I got married." She said she remembered serving Sunday dinner every week to Rosa May, part of Bodie's famed demimonde, and her girls: "They were really beautiful people and they were very nice, too. We never had any trouble with them. They were nice people."

Mrs. Annie Miller was well known for her hospitality and good food. She ran her Green Street boardinghouse clear through the Depression years. In the "quiet" years of Bodie, the 1920s and 1930s, she at times had up to forty boarders. "They didn't all room with her, but a lot of them boarded there. I don't know where they stayed . . . up in the Mansion house and different places. A lot of the guys were single that lived in town, so they ate at my grandmother's," remembers Fern Gray Tracy.

In 1911, a power plant was constructed in Jordan, near Lundy Lake. This plant brought electricity to all of Bodie, homes and mines and mills, as well as to Aurora and Hawthorne. Before a year had passed, during the brutal winter of 1911–1912, an avalanche crushed the power plant, killing all but one person. The storm that led to that avalanche was so severe that the men living in the railroad depot on the hill above Bodie decided to repair to the

This lightbulb is said to have been on for more than seventy years.

Occidental Hotel down in town until the weather improved a little, taking their horses to the Burkham barn first. They left the top of the hill at eleven o'clock in the morning, and it took them until three o'clock in the afternoon to reach the hotel.

The rescue party that reached the Jordan Plant and saved the sole survivor came from Bodie through severe snow levels and under difficult conditions. But come they did, to help others in trouble. Katie Conway Bell Adair, in an interview, recalled how her family put up the rescue party for a few days.

Katie Adair said the only survivor was a woman who was pinned under a fallen

ABOVE: *The Gray family's stamp mill in Bodie Canyon. Courtesy the Gray-Tracy Collection*

BELOW: *Mrs. Lauretta Miller Gray and her daughter, Fern Gray Tracy. Photograph by Timothy Hearsum*

Mrs. Lauretta Gray and Fern Gray Tracy

Bodie was your family, you know. Maybe because it was a sense of family. So that bond is there. —FERN GRAY TRACY *(Interview, 1988)**

"[I]n my mother's time it was pretty wild," says Mrs. Gray. She herself grew up in Bodie after the boom days had subsided, but the town was far from dead. She remembers a good-size town that had a reputation forged in its earlier, more dangerous days.

Mrs. Gray was, when I met her—she has since died—a calm and warmly polite lady of another era. When she spoke of Bodie, she lit up with the warmth of bygone times. Mrs. Gray was born Lauretta Miller in Bodie in July 1901. Her father, William Miller, a miner, was Canadian by birth, and her mother, Annie Currie, was a Minnesota native whose family came to the Mono Lake area in 1885.[†]

In time, Miss Lauretta Miller met and married a miner named Ed Gray. She and Mr. Gray began to raise a family. Their first son, Larry, was born in his grandmother's Occidental Hotel. Later they had a daughter named Fern, now Fern Gray Tracy. Although Mrs. Gray said Fern's younger sister was the last baby born in Bodie, Fern herself was not born in Bodie, but in Yerington, after a harrowing horse and wagon ride. "We had to go down . . . across the flat to Bridgeport. We had a runaway. And the guy that was driving got so scared! He thought I was going to have the baby before we got there! . . . We stayed with the sheriff and his wife [in Bridgeport]. And then we went on to Wellington, Nevada, and took a car from there and went into Yerington . . . we had quite a trip." Fern arrived safe and sound.

Although Fern doesn't recall any good ghost stories from a childhood in a town so famous for dastardly deeds, she does recall one spooky story her

---

* Mrs. Gray and Mrs. Tracy were interviewed once in 1994 by the author and once in 1988 by Frank Lortie, a historian employed by the Department of Parks and Recreation (Oral History Collection, Bodie State Historic Park, Sierra Division, Department of Parks and Recreation). This interview is an amalgamation of these two interviews.

† The Curries' days in Mono Basin are recounted in Margaret Calhoun's *Pioneers of Mono Basin* (1984).

father told them. "Across from our house was a mine tunnel from an old mine. My dad told me a tale of a miner with a white mule. The miner and the mule died tragically in the mine, and if anyone approaches the mine, the white mule will appear to keep them out. As a kid, it scared the daylights out of me and I never went near the mine. I suspect that's why he told me the tale—to keep me out. It worked. My great-nieces always get excited and want to check it out, but my aversion is still with me."

In 1942 the Grays left for the San Francisco Bay Area, but didn't sell their land, their claims, or their stamp mill, down the road from Bodie. Her father built that stamp mill, and the family owns it still. When asked what she thought about plans one mining company had a few years earlier to start an open-pit mine on the far side of Bodie Bluff, Mrs. Tracy said, "Boy, I tell you my back went right up." Virtually all the Bodieites echoed this sentiment.

The Grays' house is the last house on the left on the way out of town toward Aurora, up on the hill. It had a circular driveway at one time.

wall between her dead husband and the family dog. After the woman was rescued, they brought her to the Conways' house, which was not in Bodie, but closer to Lee Vining. Katie Adair's mother cooked for the rescue party and packed lunches for them to take.

"She'd have bread raising on top of the piano and every bureau in the place. . . . [S]ome of the men would stay up and play cards, keep the fire going, then . . . they'd change off." They taught the men to determine when the bread was ready for baking so they could have bread for their lunch.

"There was a woman that was living with a miner up there (Lundy). She happened to be an R.N. (Registered Nurse). She came down on her horse and said '[C]an I help you?' My mother said, '[Y]ou certainly can; I'm anything but a nurse,' so she stayed with my mother to take care of this lady 'cause her leg was gangrenous from lying near her [dead] husband. . . .

"Some 'nice' ladies called [my mother] up and said they were ashamed they were taking this lady into her house with her daughters. My mother got furious and said, 'I haven't seen any of you good churchgoers come over here to offer your assistance to me!' She was furious! This woman stayed there and helped her." The woman and Katie's father built the monument to honor the people who died. Her dad built the fence; the woman engraved the markers.

The injured woman was getting along well at the Conway house. When the Bodie doctor came to check on her, he told Katie's family they were doing just fine in caring for her. Just the same, the woman's wealthy uncle had her taken to Oakland, where doctors amputated her leg. Katie Adair said it was too bad, because she was coming along all right at their house.

LEAVING BODIE

When people left Bodie, it was no small chore to relocate. Even as late as 1911, there was a lot of open territory between Bodie and any sizable city. Jack Bryson and his father left Bodie for San Pedro in southern California in 1911. They took a one-horse, two-wheeled cart from Hawthorne to Sac-

Katie Conway Bell Adair

In October 1987, Mrs. Katie Conway Bell Adair* was interviewed by Bodie SHP volunteer Diana Mapstead. Mrs. Adair had vivid memories of Bodie and the Lee Vining area. Mrs. Adair's father was a friend of J. S. Cain's. Sometimes her father brought her with him to Bodie for a visit. On those occasions, he told her mother to braid her hair really tightly, because they'd spend the night there and he didn't know how to fix her hair the next day.

"He'd take me by the hand and I used to go into all the old saloons and everything with him. People would always say hello to me. So then, when I was growing up, a lady came and said, 'I'd like to go uptown, but I'm scared with all those old miners.' So I said, 'OK, I'll walk with you,' and every one of those old miners would get up, and tip their hat and say, 'Good morning, Miss Katie,' or whatever time of day it was. This lady said, 'I can't believe it!'

"Even after I was married and had children, a lot of them moved down here 'cause the mining's closed, and they'd always know me. A lot of the ladies down here [Lee Vining] would say, 'Katie, you sure have a variety of friends!' I did and I'd say, 'Yeah, and every one would give me the shirt off their backs for me,' and they would, too. . . . I never heard them swear or anything around women that didn't wish it. I used to say if you'll always be a lady then you'll always be treated as one."

Mrs. Adair also remembered some of the less proper women of Bodie: "[S]ome of them married the men there and made good wives for them . . . and they were nice ladies. . . . [O]ne might be [alive], I don't know what happened to her when she left Lee Vining."

---

* Katie Adair was not related to the Bodie Bells.

ramento. "We'd go 50 miles a day and lived on the land. We'd fish, my dad was a great fisherman. Or we'd shoot rabbits. We did not worry about food." They sold the horse and carriage in Sacramento and took the train from there. "I used to think, 'I wonder if the Pacific Ocean, when I get down to San Pedro, will be as big as that slump pond.' Then when I got back up here (to Bodie), I see how small it is!"

FURTHER CHANGES

In 1912, Wells Fargo shut its Bodie office and discontinued its transport service from the town. After this, gold was taken by automobile to Reno, but without the burly Wells Fargo "messengers" that traditionally guarded gold shipments, the miners were on their own. Although they were nervous transporting their "color," there were no major mishaps.

In 1915, J. S. Cain challenged the Standard Company in court, claiming they had been working on his claim. The court found that it was true: The Standard's underground operations had gone over the line. In settlement, Cain won control of all the Standard's holdings in Bodie, closing them down and making him the virtual landlord of the town and mines. After the Standard Mine closed, Bodie's population dropped further, and more of those who remained became "leasers," miners who independently leased portions of other people's mine properties.

In September 1915, the *Saturday Evening Post* ran an article (Van Loan 1915) about Bodie, part of a series entitled "Ghost Cities of the West." The article significantly exaggerated Bodie's fall from grace, and may have been the first source of the idea that she was a ghost town full of treasures for the taking. The author claimed there were only a few old men living there when he visited, yet there were actually a few hundred people in town—the 1910 census places the population at 698—with active, ongoing mining operations on the bluff and businesses operating in town. The author painted a picture of utter desolation in the high desert, and asserted that Bodie was dealt "her deathblow in the winter of 1884, when the Standard pulled out her pumps and allowed the lower workings of the district to fill up with water. And the water is there yet."[2] But Bodie was not dead.

In all the gold-laden days of Bodie's glory there was never a bank heist. The Wells Fargo messengers and vigilant bank clerks (recalling those who actually slept in the vault in the boom time) made sure of that. Then in September 1916, Bodie's bank was finally robbed. In fact, "burglarized" is the legal term, because the thieves broke into the bank at night, getting away with $2,000 cash and some gold and valuables stored in the vault. Had the clerks been bedded down in the vault as they were in the boom time, the thieves would have been nabbed. But those days were past, and the culprits were never caught.

By 1917, between the reduced population and the conversion to electricity, Bodie's demand for wood had greatly diminished, and the need for a wood transport railroad along with it. It was determined that the materials used in the railroad could be better used somewhere else—in the war effort, for example. So the tracks were torn up and sold, along with the engines.

FROM LIFE INTO LEGEND

Between 1915 and 1940 several articles were written about Bodie in popular publications, most of them by people who had worked or lived in Bodie for a period of time. All focused on Bodie's bawdy past and portrayed town life in a bemused, idealized way. Several of them furthered the idea that Bodie had been left full of goods and was just sitting in the desert sun. More than a few of the articles from those days are less than factual; one even switched the identities of the victim and villain in the Treloar-DeRoche murder. Still, they served to popularize the notion of an abandoned ghost town.

In 1925, C. C. Keely wrote a more factual article describing Bodie as it was long after the boom, but before the fire of 1932 that would render it as it is today:

Situated above the town on the sloping hillside were the mines. . . . At many of the entrances sheet iron buildings had been erected and poised on top of each was a big bull wheel which once turned busily hauling out the ore cars. Everything was now motionless and silent. . . . [T]he road led directly into this wide thoroughfare I had observed. . . . The buildings, all commercial ones, were built side by side and nearly every other one was a saloon. I counted 16 on the shady side of the street before departing. . . . As I drove slowly down the three blocks of the street I noticed near the end what appeared to be a hotel. I was rather startled by seeing a man lounging on the porch for he seemed so out of place in this lifeless surrounding. . . . He very kindly offered to show me the town but before starting out returned within his establishment to get Mr. Cain, the man who, as I soon learned, owned Bodie. Mr. Cain was a refined little old gentleman and wore a stiff collar and blue serge suit. . . . His grey hair and his rather sorrowful eyes showed that he was living in the past. His life was now only the imagination of what it had been. . . . [I]t was a real little bank that we stopped at. The vault was there and fastly shut. Behind the grilling stood a pair of balances for . . . gold dust was the principal medium of currency deposited . . . the scrap baskets had papers in them from the day the place closed which was in 1907.

. . . There was a Chinatown. . . . I found chop sticks, funny shoes, odd clothing and queer pipes. I climbed a little ladder . . . and I found in the loft a row of six or eight . . . double tiers. . . . I am anxious to return to Bodie. . . . One must see it to understand what it is like.

Bodie was still populated by miners and a few families. The unused dwellings greatly amused the children in the town, and although they were told to keep out of them, most sneaked in here and there when they could. And all knew the aforementioned Mr. Cain well.

Bodie was still a town where no one locked their doors. Never had. The Bodieites say this is the root of the "myth" that folks up and left the town, leaving possessions behind. As Marjorie Dolan Voss put it, "[P]lease stop that story about folks leaving everything in their houses: it's pure nonsense!"

Most Bodieites took their possessions with them when they left for good. But many kept their Bodie houses as vacation homes and left possessions and supplies in stock for their next visit. Unfortunately, the Bodie custom of unlocked doors worked against them when tourists found houses full of furniture and clothing. Believing the tales of abandonment, they helped themselves to what they wanted. When families returned, they found their homes had been gone through and things were missing. Thus those final years in Bodie were very discouraging for many former citizens.

This 1927 exchange of letters between J. S. Cain and a former Bodieite was unfortunately not unique:

Dear Mr. Cain,

This will introduce Mr. T. H. Walker, who is vacationing in your locality this year.

Mr. Walker has kindly consented to ascertain what the possibilities might be to have our goods, which have been stored with you for several years, shipped to Los Angeles. We have not been in communication with Bodie for so long that we have no idea what the transportation facilities are at present and any advice you could give Mr. Walker regarding the best procedure, cost of shipment, etc., would be very greatly appreciated by us.

We would only be interested in the boxes which are marked "Mrs. E. R. Brooks." Mother and I are living in Los Angeles at the present time. . . .

We are confident that you will do all you can to help us out about our goods, Mr. Cain, and we would be very glad to hear from you at any time. . . .

With best personal regards and assuring you of our appreciation, I am

Sincerely,

G. Brooks

J. S. Cain's reply was as follows:

> I received your letter of June —th, delivered to me by Mr. T. H. Walker yesterday. I am glad to hear you and your Mother are well and nicely located.
>
> I regret to tell you all the boxes containing your household goods have been broken open and all the material has been taken by the auto travelers, many cars come here daily and they take anything they can haul away in their machines. I have not been here out [*sic*] one half of my time in the past three years and the Tourists have cleaned up the place. I looked over the place this date and every box of your Fathers had been opened, also trunks and boxes of Alice and Julia Boone have been broken open, Alice and Julia was [*sic*] here June 27th, Their trunks and boxes were stored in the same place. . . .
>
> I wish to be remembered to Mr. and Mrs. Walker and party who registered in the Tourist book in my office, we are pleased to have all the nice people on the register and I hope we may have the pleasure of a visit from you and your Mother very soon.
>
> Very truly yours,
>
> J. S. Cain (1924)

PROHIBITION, BODIE STYLE

In the 1920s, Prohibition was the law in Bodie, as elsewhere. But as one can easily imagine, it was not a popular cause in Bodie, and the local citizenry smoothly worked around its regulations. Arena Bell Lewis, a native Bodieite, recalls that during Prohibition several families that are respectable today had speakeasies in Bodie. "Some of them referred to their establishments as 'ice cream parlors,'" she says.

Mrs. Gray agrees, saying that when Prohibition was the law, Bodie saloons went "underground," so to speak: "[W]e had about six saloons and they all were illegal. . . . Then they had what they call a hash house. . . . Every Saturday and Sunday they would go in there and boy-oh-boy, they got their liquor there . . . they'd have a heck of a good time."

Inspectors were circumvented by the Bodie storage system. Up in the cemetery were several headstones that featured screwed-on metal plates.[3] They simply unscrewed the plates and stored their whiskey therein. Government officials never tumbled to this. But after the repeal of Prohibition, Bodie people seemed much less attentive to their dear departed in the cemetery than before.

In the later years, when most of the saloons were silent and Chinatown was merely a memory, one saloon kept the social life alive, and it was run by the last Chinese resident of Bodie, Sam Leon.

Former Bodieite Gordon Bell remembers Sam well. "Sam Leon was a Chinese man. He ran the little beer joint up there . . . [he] had the U.S. Hotel . . . before it burned down in '32. Then he came back . . . and had a little bar . . . when the Roseklip was running in the '30's . . . the Dechambeau Hotel and those other buildings . . . weren't open."

Sam's lunch counter did a brisk business, even during Prohibition (it is rumored that the Temperance Order didn't rule there), but it was not a family joint. Children were not to intrude when their family members were there, and it was rumored that upstairs in Sam Leon's establishment worked the last of Bodie's "naughty girls." Gunnar "Pete" Peterson remembered "Bodie Sam," as the locals called him. "He used to bring in a couple of women on the weekends for the bachelors!" Peterson claimed. Many of the workers ate at Sam Leon's or at home unless they went into Bridgeport or Lee Vining. "We used to go to Lee Vining because the road was so much better back then," Peterson recalled.

Jack Robson does not know if there was still a "lady of the evening" working in Bodie, as some claim, but he recalls his uncle's social life involved Sam

Sam Leon and friend in front of Sam's Bodie Café. Courtesy the Bell Collection

Leon's bar. It was not a scene to which he was privy: "I was strictly informed to never go in there. . . . I went in there just once . . . and he got madder than hell and about kicked me out."[4]

WISE GUYS IN BODIE

Bodie did not escape the gangster era untouched, either. Tales of Pretty Boy Floyd and John Dillinger visiting Bodie have circulated for years. Pat and Ed Goodwin say Pretty Boy Floyd (or was it John Dillinger?) passed through Bodie one night for gas. Ed Goodwin recalls the story of when the big-time gangster appeared: "Dillinger came through . . . they were on the run. . . . My uncle pumped gas for him. . . . He saw the guns in the back seat when he was pumping the gas." Ed's uncle was Victor "Vic" Cain. Other Bodieites recall Vic Cain's brush with infamy that day as well. The carload of people did not stay in town, and, just as in the Bodie of old, no one in the town asked any questions.

Another time, when it was rumored that Pretty Boy Floyd was on the approach, the miners became nervous about their gold. According to Robert Sprague, nephew-in-law of miner Marion Raab, Marion poured one enormous bar of gold using all he had milled, one bar too heavy for anyone or even two or four people to take away, even at gunpoint.

Dr. John H. "Jack" Robson

Jack Robson is a lively man with a devilish sense of humor. A lifelong scientist by nature and profession, he is at once both inquisitive and informative. Jack spent several of his early summers in Bodie, living with and helping his uncle, Marion Raab, and his grandfather Philip Raab (who was there off and on) in their mining and milling activities. The Raabs were here from the late 1920s until about 1935, when Marion Raab had to leave for health reasons. Marion Raab was a "leaser," a miner who leased property from J. S. Cain, mining and milling the gold he found. In the 1930s, they lived in what is now called Dr. Street's house, on Green Street across from Mrs. Miller's boardinghouse. Dr. Street had been the company doctor for Treadwell-Yukon.

On his first visit to Bodie in decades, the rangers let him in to take a look around his old house. "When I was a kid, most of these houses had belongings in them," he says, gesturing around. Although Jack, like other children in Bodie, was forbidden to enter the empty houses, he, probably also like other children in Bodie, liked to play in them on the sly. He had a "secret clubhouse" up in the attic of the house behind their house. "I even had an old kerosene lamp I found in it. I'd go up there and light the thing . . . gives me the shivers now to think I did that!"

As Jack came into his old house, he was impressed by the changes that had taken place. The wooden porch he remembered was gone; there was no sign of the partitions they had rigged up for rooms. In fact, he was amazed to view with adult eyes the house that was home to him, his uncle, and three others. "I just can't believe it shrunk so much!"

The whole household took their dinners at Mrs. Miller's boardinghouse across the road, but otherwise ate in the kitchen, "and it was ten times bigger than this," Jack swears. He gazed up the hillside and recalled swimming in the water supply tank until they were chased out. "We were warned to stay out of it, you know, so naturally we didn't."

He recalls a pantry where they kept "a big bunch of burlap hanging down over the cupboard and then there was something keeping water dripping down it all day to keep things cool. . . . The ice man didn't come by very often."

He also recalls making use of that pantry the day he got into a fight with a Kuzedika boy, "a great horrendous brute of a kid." When the youngsters were having a good-natured rock fight, Jack clobbered him with a rock and ended up hiding in that pantry: "I wasn't about to fight a kid twice my size."

Dr. John "Jack" Robson in his Bodie summer home of childhood

But when his Uncle Marion found him, he spent an hour teaching Jack "how to keep that left out there" and then insisted that Jack fight the boy.

"Took me in the middle of the road. The whole damn town was here . . . [even] all the men that were eating dinner over there [at Mrs. Miller's boardinghouse] . . . [e]very kid in town. . . . I was definitely the favorite because I was so much smaller than he was." And as a rare stroke of fortune afforded, Jack won. "That was a great night," he recalls with pleasure.

Even as late as the 1930s miners worried about the possibility of marauders coming for their gold. The precautions taken when it was rumored that a gangster was in the area have already been described, but Jack was given a job as a lookout for the miners, a job he's never forgotten and probably never improved on. But he came by it after a Tom Sawyer-esque adventure.

It seems the wife of his uncle's partner, Mrs. Murdock, was in charge of young Jack when the miners went off to work every day. She would have Jack cleaning and working all day, leaving her free to relax, as he recalls. At some point he became fed up. He packed up a couple of cans of beans and his coat and set off on foot, determined to walk back to Los Angeles rather than be this woman's slave. He did not think of taking water, nor did he let anyone know of his plans. He headed down the Cottonwood Creek road toward Lee Vining. Toward sunset he reached the Scanavino Goat Ranch, about ten miles from Bodie. As some cowboys approached him, he hid in the sagebrush, but was not fast enough for these experienced ranch hands. They nabbed him and brought him to the ranch, where they called Bodie and asked if anyone was missing a boy. When it turned out that someone was indeed missing a lad, they brought him back to his Uncle Marion in Bodie, where Jack confessed all, including his reasons for leaving.

Marion gave him a new job the very next morning. Jack was their full-time sentry, perched up on top of the roof of the mill. "That's where I sat with my .22," he points out, "to keep the robbers away while we were making a big milling. . . . I could see all the entrances to the town. . . . [I]t was my job and they were very serious about it, really."

Did anybody ever come? "No. I never did anything but sit there with a .22 rifle across my lap." But he did feel important. And it was much better than cleaning for Mrs. Murdock.

Still, Jack says, the miners were cautious and kept themselves well armed when milling their gold. "[T]he word had gotten out there was a big milling going on at Bodie and everybody in Bridgeport and Lee Vining knew it was coming up. . . . [E]verybody had a gun in there. I'll bet there were . . . 20 guns around that mill."

For Jack the visit to Bodie was very rewarding. After stopping by his old haunts, he capped off his tour with the fulfillment of a childhood desire: With the ranger, he entered the back of the Boone store and got to look around. It had been strictly off-limits to children when he lived there. "Old man Cain had his desk right there," he claimed. "We were never allowed in then. He kept a roll of pennies [that were fused together in the fire] on his desk." When the ranger asked if it satisfied some longtime yearning to wander through J. S. Cain's office, Jack smiled softly. "Yes. Yes, it does!"

Pretty Boy Floyd never came up to the mill to take their gold; however, he did put in a peaceful appearance outside the Bell residence. "They came into town one night and talked with my dad for over an hour in front of the house," Bob Bell says. Floyd and his friends were hiding out in the canyon near Aurora. Later, when Bob's family read about the shoot-out in which the federal police killed Floyd and his gang, his dad recognized the woman who had been with them in Bodie, from the newspaper picture.

Despite the worries of local miners, no gangsters ever actually tried to steal their gold. Pat Goodwin points out, "[Y]ou'd be stupid to try to . . . steal anything from Bodie because you could be stopped at either end. They could put roadblocks at either end. You couldn't get out of there."

TREADWELL-YUKON AND HOMESTAKE DAYS

From 1928 through 1931, the Treadwell-Yukon Company conducted the first systematic exploration of the Bodie Mining District. In 1929, with the Homestake Company, it commenced work. The companies were interested in digging an open-pit mine, milling the ore remaining in Bodie dumps, and pumping out the water to work in the Red Cloud and Noonday mines. However, metallurgical problems slowed their progress. Then, once the Red Cloud and Noonday mines were pumped free of water, the modern miners had just as much trouble as the old-timers in following ore bodies in Bodie's uniquely complicated geologic structure. In 1932 the companies reevaluated their interest in Bodie, each company assigning one of its geologists to the case.

The Treadwell-Yukon geologist voted to keep at it. He felt the company could solve the metallurgical problems and that there was worthy ore in the southern mines. However, the Homestake geologist disagreed, and in the end his opinion prevailed. The two companies pulled out of Bodie and took their machinery elsewhere (Loose [1979] 1989). The Homestake did maintain its interest in Bodie for many years and to the turn of the twenty-first century was not entirely unassociated with claims in the area.[5]

THE GREAT FIRE

In June 1932, Wynne Bennet Taylor turned four years old and was given a birthday party. One of the guests was a little boy nowadays referred to as "Bodie Bill." Billy Godward was the son of a restaurant owner and a saloon-keeper. When he arrived at the party and found that instead of the traditional cake and ice cream they were to have Jell-O, he was put out and stomped off.

Bodie Bill, the little fellow who "done it" in June 1932. Courtesy of California State Parks, 2002

No one worried about serious harm coming to their kids in Bodie in those days; it was unthinkable. But Bodie Bill could find trouble on his own.

Arena Bell Lewis remembers Billy Godward: "I remember he always had a pacifier in his mouth, even when he set the fire." Bob Bell, Arena Bell Lewis's brother, also remembers him. "He was a real firebug. My mother caught him lighting fires just outside our house [before the big fire]. She caught him, cuffed him and stamped out the fire."

His mother also discovered the big fire when it happened. She saw smoke coming out of an abandoned building that morning and went to investigate. There she found Billy. Bob says, "He had lit a piece of dangling wallpaper, and was still standing there watching. He said, 'See what I done?' "

Mrs. Bell grabbed the child and ran for help. It was a windy day, which worked to fan the flames. Various other elements combined to make it the worst day in Bodie's history.

When the fire hoses were attached and turned on, only mud and stones came out: The water system grates had not been cleaned in ages. People at first tried to save what they could, then watched helplessly as the town burned. Bob Bell says the whole place burned within about an hour. The fire burned roughly 90 percent of Bodie. "The wind changed direction and that's what saved the rest of the town," he says.

Every Bodieite who was there vividly remembers the great fire. The memory still makes every one of them shudder. Arena Bell Lewis describes it: "That was such a horrible, horrible feeling. . . . There was an old saloon with old glasses, a player piano, old mirrors and things. . . . Even at 12, I realized that valuable old antiques were burning. . . . After the fire, in Old Chinatown we found old spoons and little bottles that may have been from the opium businesses there."

The day that Bodie was almost destroyed. Courtesy the Dolan-Voss Collection

They say a change in the wind's direction is what saved Bodie. Courtesy the Bell Collection

All the Bodieites remember the day of the fire with horror. Courtesy the Gray-Tracy Collection

Fern Gray Tracy recalls the panicked effort to fight the fire. "When the fire came, I went to my grandma's boarding house [Annie Miller's on Green Street]. My brother ran off to help fight the fire. I got up on the roof—I was eight—and my grandma handed me buckets of water which I threw on the roof. . . . Sparks were flying all over . . . all the men came out of the mine. . . . Mom, of course, was down home. She didn't know where anyone was. . . . She came first to my grandmother's to find both of us. And one of us was of course missing . . . then she had to strike out and see if she could find him. She found him later, he was helping locate Mrs. Riley's false teeth. She had run out without them. Our mother was mad." Mrs. Gray, Fern's mother, concurred. She was indeed mad to have had such a scare.

"Everybody got really sick . . . afterwards. The men were all so tired, you know," says Fern. "[I]t's funny . . . because to this day, [when] friends of mine . . . hear a fire engine, they want to run and see the fire. And I just want to go in the other direction. I don't want any part of any fire."

Although Jack Robson was not in town for the fire, he remembers coming back to Bodie afterward and picking through the ashes. "[A] bunch of us kids built a little screen and we took it into a couple of burned out . . . buildings. . . . [W]e shoveled ashes up and we found coins and sometimes rolls of coins. And Old Man Cain would spot us and come out and chase us away."

There had been a smaller fire the previous winter at which time the clogged traps of the water supply were discovered. No one had cleaned them out in the meanwhile. In a stroke of irony, J. S. Cain, who was famous for his desire to preserve Bodie and his faith in her eventual comeback, was the man selling Bodieites water in those days. The maintenance of the reservoirs and fire hydrants, and cleaning the traps, had been his responsibility.

J. S. CAIN

J. S. Cain and his daughter-in-law, Ella Cain (Mrs. Victor Cain), purchased some of the houses left by people who departed after the fire. "Ella [Cain] bought most of the houses in Bodie. . . . She bought them for taxes when people didn't pay their taxes . . . not that she got money out of it, but she wanted to save them," says Pat Goodwin.

J. S. Cain had remained in the town year-round from the height of Bodie's boom days and hung on, even after the disastrous fire. He still dressed in a suit and opened his Bodie Bank each day, despite the departure of friend after friend and family members for more populated places.

The Bodieites interviewed for this book all remember him through a child's eyes as friendly, but stiff, a far cry from the "Bad Men of Bodie"

Vic Cain skiing off the store roof at Main and Green streets. (But is he wearing a suit?!)
Courtesy the Gray-Tracy Collection

who had been Cain's fellow citizens in the earlier days. "He was always very proper, very dignified, very much the banker," said Pat Goodwin, and the others agree. He was sort of the mayor because he "practically owned all of Bodie."

"You know, I never saw him out of a suit. He never wore overalls like everyone else. His son Vic was the same way," says Bob Bell. "I even saw Vic saw wood in a suit!" Most of the townsmen wore overalls unless they were in their Sunday best.

Bob Bell's cousin Gordon concurs that J. S. Cain was very dignified and basically all right to kids, but "He used to have [what] you'd call . . . a bad habit. You'd talk to him and he'd reach over and . . . pinch you. He'd give you a pinch and say, 'Gee, that's pretty thin material, isn't it?!'"

Eventually even J. S. Cain's wife left Bodie to live in San Francisco with their daughter and son-in-law, but J. S. Cain remained to await the new Bodie boom. Finally he, too, had to repair to the city by the bay. He died in 1938 in San Francisco.

DEPRESSION-ERA BODIE

By the time of the Great Depression in the 1930s, there was no longer any Bodie newspaper. The citizens relied on the Bridgeport paper for their news. When the fire of 1932 devastated the town's buildings, it devastated the last of the citizenry as well. Some people stayed as long as there were working mines, but several families left. Most of the Bodieites interviewed for this book spent most or all of their childhood in Bodie, at least partly during the Depression years. Bodie was a good place to live during the Depression because although life there always involved hard work, it seems to those who were there that life did not worsen or even change much at all during one of the United States' most difficult economic periods.

As Fern Gray Tracy remembers it, "Life was pretty much the same. I can remember we'd buy food in Bridgeport at Victor Cain's general store. Then my dad would pay his bills with gold he'd mined out of his claim. We never went hungry during the Depression."

Fern remembers Bodie as a small town of about 300 to 400 people when she was growing up. She says it was a town of people who knew each other but were still separate units. "Families did stick to themselves and didn't socialize tremendous amounts. People kept to themselves but they were friendly."

Mrs. Gray added, "Our house was the friendly house, the visiting house. Folks would come to our house in the evenings." Other Bodieites also mentioned fun evenings spent at the Grays.

Growing up, Bodie kids didn't have city-style fineries. The town that had once offered such luxuries as fresh East Coast oysters and hotel dining-room menus fit for royalty was now a tiny rural enclave. They didn't have the newest modern conveniences such as bathroom plumbing in every house and consistent hot water. There was no doctor, and there were no police, just Constable Ed Gray, Mrs. Lauretta Gray's husband. He had the authority to arrest people, if necessary. According to Arena Bell Lewis, Dr. Street, the Treadwell-Yukon doctor, was possibly the last doctor in Bodie, practicing medicine there in the 1920s.

But by and large police and doctors were not needed. Everyone was astoundingly healthy, and no one could tell if it was flu season, probably due to the isolation from large populations. People were also careful to stay healthy: Memories of smallpox in the late 1890s and the flu epidemic of 1917–1918 lingered in the older citizens.

Fern remembers the ever-present mining and milling operations: "[W]hen the Standard Mill was running, you could feel the noise in the ground." The noise did not bother them; they were used to it.

Although Pat Goodwin does not personally recall the noise of the stamp mill (other Bodieites remember how it shook the ground continually), she does recall that once in the middle of the night "the stamp mill quit running. . . . It stopped for some reason and everybody woke up!"

Gordon Bell remembers the mill noise. "The mill ran 24 hours and day, seven days a week. Everyone worked eight hours a day, seven days a week. If you wanted time off, you'd trade with another two operators and they'd work twelve-hour shifts. Then, if they wanted a day or two off, you'd reciprocate."

The amazing variety of goods available in Bodie's earlier times had shrunk. Now residents had to plan ahead for their provisions. The Grays kept chickens. Fern's father built a root cellar where they kept root vegetables and fruits under layers of sawdust. This let them have fresh vegetables and fruits all winter long.

Marjorie Dolan Bell Voss didn't have a root cellar the way the Grays did, so they ate "a lot of canned goods from Bridgeport. Supplies would come in on the stage."

Sage hens were fairly plentiful in Bodie. They even sat on the Grays' porch, Fern Gray Tracy says. "But the men had a hard time getting them on hunting expeditions. They'd fly up and flutter with such a noise, it would scare the hunter and he'd lose his chance. We ate a lot of sage hen, and some deer."

Sage hens, a Depression-era dinnertime staple, continue to live in Bodie.

Gordon Bell, the real *Bad Man from Bodie!*

Gordon Bell

Everybody got along with everybody. People would help other people.

—GORDON BELL, Bodieite

Gordon Bell remembers having tea with his Aunt Louisa in her home when tourists opened the door and came on in. "We'd be sitting there, curtains on the windows and flowers outside and smoke coming out of the chimney. . . . [P]eople would just open the door and walk right in. 'Oh, I didn't know anybody was in here—I thought these houses were all deserted!'" It happened all the time once folks had heard Bodie was a ghost town. But it wasn't.

Gordon Bell lived in Bodie as a young child. Even after his family moved out of Bodie, he visited often and returned there as a young man, living in the Bell House and working in the mines for four years. Gordon is a first cousin of Bob Bell and Arena Bell Lewis. Their fathers were brothers. Gordon's father worked as a teamster for Cecil Burkham. His mother was a McKenzie, the daughter of the owner of the Bodie Brewery. Gordon's grandmother was Charity Wells, sister of J. S. Cain's wife, Martha Delilah Wells. The Wells sisters came from Genoa, Nevada. Gordon's grandfather Lester E. Bell and J. S. Cain both came from Canada.

For Gordon, Bodie isn't an interesting historical relic, it's a place where a lot of family history took place. For example, the Burkham House was owned by Cecil Burkham, who was noted for bringing the first horseless transport into Bodie. However, Mr. Burkham gave that house to Gordon's father and mother as a wedding gift, and they lived there for some time.

Ella Cain had no mummified head that Gordon knew of (mentioned by another former Bodieite), "but she did have a stuffed magpie." This magpie had been a wild one that attached itself to Gordon's cousin Arena. When alive, the bird had liked to talk and would go up to the firehouse. "It would sit up there and whistle at the dogs. Dogs would go crazy!" Gordon chuckled. It was found dead, presumed to have electrocuted itself while constructing a nest in the Hydro Building. Ella had it stuffed and kept it in her house.

Gordon also recalls that his father had many mementos of Bodie. "My father used to have a big shoe box full of gold samples, you know; quartz with all the wire gold through it and stuff. People would come over, [he'd say] 'This piece came from this place,' and so on. Pretty soon you end up with an empty shoe box!"

To this day, Gordon swears by one standard Bodie remedy: "squaw tea," made from a local plant. It's supposed to help respiratory problems and colds very well. One friend of his asks him to send it for his daughter's asthma because it's the only thing that's given her relief. "It's kind of bitter, makes a dark tea," says Gordon. But he swears it works.

"We had no running water until I was almost grown," says Fern Gray Tracy. "Instead, we had a well that was some distance from the house. We used to bathe once a week." Baths were a tub-and-kettle affair in the kitchen, a far cry from the whirlpool bathtub of today. "I'm not sure how often we'd wash our hair, but I do recall Mother used to use Mono Lake water or salts from Mono Lake on our hair and it made the hair very soft. Some people thought Mono Lake had therapeutic value and would soak in it.

"We had an outhouse toilet, but no covered walkway to it like the Cain family did. In winter, the men would carve a trail through the snow to the outhouse. If the snow was too bad, we'd stick to using the chamber pots until they'd carved the trail," says Fern.

Ed Goodwin remembers outhouses as well. "Imagine going out there in the wintertime," says Ed. "We used to go to dances in Bodie and toilets would be outside and I remember the girls had to sweep the snow off the seats before they could use it. Just think how cold it was!"

Gordon Bell was amused after his reference to a "thundermug" brought a bewildered response.[6] He recollects that "most everybody had hot water . . . everybody had bathrooms," but this did not necessarily include a commode. "Henry and Virginia Klipstein . . . in the house right next to where the Standard Mill is now, they had indoor plumbing . . . there were a few houses [that] had indoor plumbing." The Cain House eventually had indoor plumbing, but most folks were still using outhouses. "We had two. . . . In the wintertime there was an indoor privy, . . . located at the far end of the wood-shed, and . . . there was an outdoor privy. . . . In the summertime, you'd use the outdoor privy and in the winter you'd use the one in back of the wood-shed so you didn't have to go outside." Others in winter had to make do with thundermugs indoors.

Hot baths were the norm year-round: "We all had wood stoves . . . they all had the coils in them with the tank right next to the stove, so you'd have hot water. We had . . . those galvanized tubs, like washtubs . . . they were oblong. . . . At the mill on the hill they had showers."

SCHOOL DAYS

The Bodie school was still in session during the 1920s and 1930s. School days were filled with the customary chores and lessons in the schoolhouse on Green Street, where, the Bodieites say, they received a better education than many of their counterparts in the big cities. Stuart and Sadie Cain recalled

The Bodie schoolhouse with just a few pupils. Courtesy the Gray-Tracy Collection

that grades three through six were in the room visitors peer into today; the primary grades were in a large room behind that room, and grades seven through nine were upstairs.

In general, the children attended school in Bodie until they were high school age, when they would often be sent away to other relatives to go to school; there was no high school in Bodie. All the Bodieites interviewed for this book (save Mrs. Gray) recall Ella Cain as their schoolteacher for at least some of their school days. All say she was a marvelous teacher who was fairly relaxed about behavior, but did not allow her boundaries to be crossed.

Arena Bell Lewis remembers Ella Cain's dog, which came with her to school every day; it had been a shepherd's dog. Shepherds would occasionally bring their sheep through the Bodie area; they would camp just below town. When they were in the area and made stew, Arena and Vic Cain (J. S. Cain's grandson) would join them for a meal. "Once the sheepherders' dogs had puppies. There was one longhaired red pup we played with too much for the herders to train, so they gave it to us. Ella Cain kept it, and the dog went with her every day to school. It would keep still in class, then play with the kids at recess."

Fern Gray Tracy remembers Ella Cain as a good teacher, understanding but firm, who occasionally rapped the knuckles of kids who acted up. Generally, these were boys, according to Fern. "My classroom in the eighth grade had five kids," she recalls. In Fern's day, most children brought their lunches to school, although Fern and her brother went up the street to their grandmother's boardinghouse for lunch.

Bob Bell also had Ella Cain for his schoolteacher. "I can remember her in the classroom, sniffing the air and saying, '[D]o I smell an orange?!'" he chuckles. He also remembers Mrs. Cain sitting in back in her swivel chair, reading and rocking, slowly leaning farther and farther back until the chair broke, toppling her over backward and ending her up on her head on the floor with her legs in the air! A cherished memory for any student.

Several former Bodie scholars also recall school days sometimes brought pranks (especially at Halloween, when outhouse tipping was a grand tradition). Bob Bell says carving your initials in your desk was a big deal, as the desks show.

Arena Bell Lewis

Arena remembers well her dog-calling magpie. "It was always with me. Where I went, it went. My dad said it thought I was another magpie! It would come to my bedroom window and say 'get up Magina ["Mageena"], get up you lazy Magina!'"

Arena's magpie adopted her when she was about eight or ten years old. Mr. Scanavino, of Goat Ranch fame, had caught the magpie and given it to "Tuffy" Miller (Lauretta Miller Gray's brother, Fern Gray Tracy's uncle). "He turned it loose in Bodie and it flew all over Bodie," says Arena. It suddenly landed on her shoulder as she was on her way to school. From that moment on, it was her magpie.

"The men who hung out on the boardwalks in town taught it to swear. It would peck anyone who came near me, including my father. One day it even followed me into class and pecked the teacher. She hit it with a book. It knocked the bird out, but it came to and was all right."

The bird nested in the Hydro Building and accidentally electrocuted itself there. "The lady from the Hydro Building had it stuffed. After that, Ella Cain kept it and later took it to her Bridgeport store. I never saw it again."

Arena Bell Lewis is Bob Bell's sister and Gordon Bell's cousin. They come from a well-established Bodie family. Their father, Lester L. Bell, was an assayer/chemist for the Treadwell-Yukon Company. When Arena was a child, their family lived in Pat Reddy's house. Her grandparents lived in the Bell House, her parents lived in the Reddy House, and later, she and her own family lived in the Metzger House.

Others recall the time a young man with great engineering promise (rumored to have possibly been Bob Bell himself) devised an elaborate joke on the teacher. The teacher always started the stove fire before school so the room would be warm when the children arrived. The prankster stuffed rags in the stove chimney (this was *before* the 1932 fire), then strategically poised a bucket of cold water at the schoolhouse bell, and, for the coup de grâce, stashed a stolen pig in the storage room.

The teacher started the fire in the stove and left as usual. She returned some time later to find the schoolhouse filled with smoke. Panicked, she ran to ring the bell and summon help. With the first pull on the rope, she was doused with icy water. Sputtering and gasping, she made her way to the storage room for a towel. There, unbeknownst to anyone, the captive pig had overturned the supply of red ink on himself, and as the sopping teacher opened the door, what appeared to be a partially butchered pig rushed past her and into the street.

The former students still smile when asked about it. "We got into some trouble for that one," one says mildly. The children's assignment for the day was to capture and return the pig to its proper owner, a strictly pass/fail exercise!

WINTER IN THE 1930S

Winters, generally severe until the 1930s, brought delightfully tall drifts of snow for skiing, sleigh riding, and a good excuse for congregating at the Grays' house. Parents worked hard at preparations to keep their families well supplied through the season. Arena Bell Lewis's family went to Reno each year before the snow started, to buy peaches, pears, and other fruits and salmon for the winter. "The snow started in November and you couldn't get out until April," she says. The Bodieites kept their Christmas merry, even when it meant skiing a long distance to bring back a real Christmas tree tied to your back, which Fern Gray Tracy's father did every year.

In the winter, mail came once to three times a week by horse-drawn sleigh, when the weather allowed. If they needed something from Bridgeport, they'd call and request it, and it would come up with the mail. Arena Bell Lewis knew the driver, Charlie Fulton. "They had tall stakes in the road so they could find the road in winter. Charlie Fulton drove his horses in once a week with mail and other deliveries. The horses knew the way and I remember they would arrive with big icicles hanging from their noses."

At home, Arena and Bob's mother kept petunias in the inside windows and had canaries, a welcome sight in winter. Charlie Fulton said it was like

Bob Bell

Bobby Bell—gets up with the sun and goes to bed when he's tired, no use for clocks. He and my husband went hunting once and the car broke up on a hill. Bob said he needed sagebrush—said, "you can fix anything if you have a knife and sagebrush!" He fixed the car, all right. —FERN GRAY TRACY

Bob Bell is a magical man with a somewhat shy demeanor. His Bodie pedigree is unarguable: his grandfather was Lester E. Bell of Bodie, and his father and brother were Lester L. and Lester F. Bell, respectively. Bob was born and raised in Bodie, leaving only when World War II demanded it, and returning immediately thereafter despite employment opportunities elsewhere. You could say he was the last man out of Bodie because for several years he and his wife lived just outside the town in a trailer, long after other residents had become summertime visitors in their own homes. He has seen the town go from a small, close-knit town where the children played freely and everyone knew everyone else to a state park full of tourists from all over the world.

"I began working in the mines with my father when I was about 15 or 16 years old," he says. They started in on the old Standard shaft where Bob recalls there was a station just behind a large cave-in. There, they blasted out ore behind the cave-in, unconcerned about what had previously caved in from above. "You don't care what's over you, you care about what's under you: that's what counts," claims Bob, "because you can step out on a weak ledge and go down, down, down."

As his sister Arena said, their home was very homey; his mother had a rhubarb plant that actually grew in Bodie, and Bob is pleased to point out that the plant is there today, at the Bell house.

Bob had very fond memories of Mrs. Miller's boardinghouse and its social scene. "She used to put out some pretty good meals for a dollar a meal. . . . [A]ll those young guys . . . not family men . . . they all had to have someplace for a bunkhouse and someplace to eat, so the company had to keep a restaurant going all the time. . . . [S]ometimes they had their own, if they couldn't keep up, then they'd have a different place for them to eat or a bunkhouse or boarding house."

Bob's general friendliness and magic touch—especially with things mechanical—are appreciated by all who have met him, even by those who, like the people who were stranded on the Bodie road after hours in the summer of 1996, receive his help without ever knowing who he is or how important he has been for the ghost town of the high desert.

But asked about his precious hometown today, Bob sadly shakes his head. "Bodie is nothing but a tourist trap now."

Bob Bell, the last Bodieite to leave town.

a piece of springtime to come in and hear the birds and see the flowers. Mrs. Bell would put the birds and plants on their black chair and put blankets over them at night to keep them warm.

Mrs. Gray remembers, "In winter, everyone would ski down to the Post Office and hang out waiting for the mail. No one knew when it would arrive or even *if* it would arrive on a given day. At some point someone would say, '[W]ell, where are we going to eat tonight?' and someone would volunteer, 'I'll bring a rooster and some potatoes,' and so on. And that's how we'd do up in Bodie."

"In winter, we skied to school," says Fern. "During school recess, we would toboggan down the Green Street hill. Sometimes in winter, to get home from school, I'd put my feet, one on each of my cousin's skis, and hang on to his waist." Her family was concerned that she not attempt to ski home alone through a snowstorm. "They were all afraid I'd get lost or blow away! I was supposed to go to my grandma's boarding house instead of home if the blizzard was too bad. A couple of times my brother convinced me we could make it, but then we got in trouble when we got home. The bad patch was between the cyanide ponds and the house; that's where we could get lost."

"In the winter, kids would go for sleigh rides and such in the evenings and then come to our house for food afterwards," said Mrs. Gray. In general, the Bodieites agree, people got together and entertained themselves, singing freely, without fear of being "good enough." They often played cards as well.

Mrs. Zady Kriel remembered coming from the University of California to visit her parents, Clyde and Annis Harvey, in Bodie in the 1930s. They were living in the McMillan house. "There was . . . a lean-to pantry, uninsulated, in which a side of beef could hang all winter, frozen steaks and roasts to be cut with a hacksaw. . . . Christmas vacation, 1936, I took the Greyhound [bus] to Bridgeport and rode the postman's sled to Bodie. We had a kerosene lantern under our laprobe to keep our feet warm. I found the guest couch in the living room comfortable, what with the oil stove and several comforters, but the storm howled outside. I watched the snow blown through the keyhole six feet into the room, and round spots of frost on the wallpaper marked nailheads in the walls."

Fern Gray Tracy and pals skiing near the schoolhouse. Courtesy the Gray-Tracy Collection

Sometimes most of the town would turn out to ski together. Bob Bell is the young man on the far right with no jacket. Courtesy the Gray-Tracy Collection

Gordon Bell also remembers very clearly the fierceness of Bodie winters. "I remember one winter when a fellow who was trying to get to Bodie had a problem with his car because of the cold. There was no antifreeze then, you know. He filled his radiator with wine and made it all the way back to town!"

He remembers as a young man getting up in the morning and skiing over the house across the street to get to work at the mill. "At night, in the wintertime, if it was a real bad storm, you'd go up the hill from town on skis, and more than once in the winter, why, if you started downhill on the other side, . . . you'd miss the mill! This is a big mill, made a lot of noise . . . and it was lit up pretty good, too . . . but the wind . . . would be howling, [it'd be] snowing and you couldn't see or hear . . . and then you knew you'd missed the mill . . . and you'd have to come back up the hill."

Getting home again would be a chore as well. "I used to come off the swing shift at 11 o'clock at night. You'd look down the hill and the only light'd be on would be the one at Sam Leon's bar. I used to ski down the hill and head for Sam's bar; you'd know the light, so you'd know where you were. It'd be dark. . . . [I] almost broke my neck one night. The other operator and I . . . came off, put our skis on and started down the hill. They'd taken a Cat (plow) and gone along the bottom over the road . . . over to the old stamp mill . . . [we were] zooming down the hill and we hit where they had plowed all the old road out!"

Bob Bell says the whole town enjoyed winter. "In winter, the whole town used to come out for night skiing under the full moon," Bob recalls. "They'd build a fire up on the hill at the top of Green Street and folks would climb up and ski down." Everyone skied together, not just the children, he says.

Marjorie Dolan Bell Voss recalls that although Bodie winter life wasn't easy, everyone got by with time-honored methods. "If I wanted to go visit my

Bodie women in their "snowshoes." Courtesy the Dolan-Voss Collection

LEFT: *Marjorie Dolan Voss, popular Bodie cake baker.*
ABOVE: *A baby Bell, Marjorie Dolan Bell Voss's child, in Bodie. Courtesy the Dolan-Voss Collection*

mother-in-law I'd put on my skis and had the baby in this arm and one pole in this arm and that's how I went over to visit her."

Marjorie slipped her street shoes into her skis. She didn't wear special boots like they do today, and she used the skis more like "snow shoes" (as they were called) than skis. "We just wore plain old shoes. . . . I didn't ski either. I just walked on top . . . just to stay on top of the snow. . . . Once they got the roads plowed out you could walk, you know."

Gunnar "Pete" Peterson, the assayer for the Roseklip operation, said he and some friends made some of the horse snowshoes that are in the museum: They were not extremely successful, but necessary in the snow. He said the mail and county crews tried to keep the Mono Road plowed in winter, but they were once snowed in for a month.

SUMMER IN THE 1930S

If winters were coldness and isolation for Bodie residents in the 1930s, summers were long nights of group fun. Young people enjoyed times of kick-the-can around the Cains' house, weenie roasts in the evening, and wonderful dances in the Miners' Union Hall that went on until the sun came up.

Surprisingly, ghost stories did not figure into a Bodie childhood. The interviewees all agree there were not too many stories about Bodie ghosts, with the exceptions that some remember that there is supposed to be a ghost in the upstairs of the Cain house, and Fern Gray Tracy has her eerie tale of a phantom mule.

Marjorie Dolan Bell Voss

Marjorie Dolan Bell Voss, Pat Goodwin's sister, was not born in Bodie, although she spent her summers there growing up. Their father was an electrician who worked for a while at the famed hydroelectric plant of Bodie, at Green Creek.

After she graduated from high school, Marjorie eloped with Lester Bell, brother of Bob and Arena Bell. She was living in Lone Pine, he in Bodie. He picked her up, and they headed for Reno, but stopped in Bodie for the night and to borrow enough money from cousin Gordon Bell to get married. The car broke down on the Cottonwood Canyon road, and the Bell boys had to come out to help them. They wrapped canvas around her shoes (which she recalls were probably Oxfords or "saddle shoes"), and they hiked the mile or so into town. She slept in a bed between his sisters.

After their marriage, she and Lester Bell settled into the house that has been called the Dolan House, but is now more correctly known as the McDonnel House. They had running cold water, but no hot. The water came from Masonic Springs: "[I]t's the best there is in the world," she says. It was not a fancy life, but it was pleasant. "Everybody knew everybody else and nobody locked their doors in Bodie," she recalls. She had her first baby here, a son.

She is amazed now to remember that it was common for people to leave their babies in their cribs at home with oil stove and heater going, and go to a friend's house for the evening, coming back often to check on the baby. "Everybody did it. Imagine!" she says, marveling at the difference between then and now. The babies all did fine, though Marjorie is daunted now at the idea of having a baby in its first year in a remote town without a doctor.

Although this seems astonishing to a city dweller in the twenty-first century, Bodie was a very small town with no strangers. Anyone approaching town was heard well before he or she arrived (just like today). Nevertheless, Marjorie was always nervous about staying alone at night, which she had to do when her husband worked the night shift at the mine (they changed shifts every three weeks). "I had a girlfriend that used to say, 'Don't you hate living up so close to the graveyard?' because my house is sort of the last one. . . . I said, 'It's not the dead people I was afraid of!' Anyhow, the Bells had a big police dog and on the nights that Les worked the night shift . . . [the dog] always came and stayed with me and slept under the bed."

Marjorie cooked on a woodstove, making a cake every day because her husband had a sweet tooth, as did his brothers and his cousin Gordon. "He

[Gordon] and Bobby and his brother would come over and eat some of the cake and I'd have to make another one." She is today quite impressed she did this on a woodstove at more than 8,300 feet above sea level!

Marjorie and her husband, Daniel Voss, used to hold "Bad Man from Bodie" parties every year for the Bodieites until a few years ago, after her mother passed away. They held her mother's 89th birthday party in Bodie. The park rangers opened up the houses for them and made them feel at home in their old town for one more family get-together.

The wallpaper that is in the Dolan/McDonnel House (number 1 on the park brochure tour) today is the same paper Marjorie put up as a new wife and mother. She also says the Miller house that is open to the public, although full of interesting things, is full of very different things from what it contained when she lived in Bodie.

In the warmer season, a truck with fresh vegetables and fruits would come to town, where they'd have a sort of open market. "Then they'd come down to our house because it was a little ways out from town. Me, my brother, my cousin and some friends stole a watermelon from the truck once. We really thought that was something!" recalls Fern Gray Tracy.

There was a ballpark on the road by the willows, as Pat Goodwin recalls, "outside of town, above the willow pine. . . . We used to walk there— sometimes take our lunch and go up there for a picnic."

Ed Goodwin adds, "The miners used to play the cc's. . . . [T]hey had a cc [most likely the Civilian Conservation Corps] Crew camp up in there. . . . [A]ll these kids were from Brooklyn living up in that area and used to play the miners or whoever could get together a baseball team out there."

Pat concurs, "They had good teams up there in their day."

Summertime was also the time for the Bodie dances, a tradition that began long before the Depression era. The Bodieites all swear that Bodie had the best dances in Mono County. They held them in the Miners' Union Hall. "It would start about eight or nine and [you'd] dance until midnight. They'd stop for an hour; the orchestra would rest and everybody would go off and have a couple of drinks or eat. Annie Miller used to serve dinner sometimes. She'd serve midnight dinner and then you'd come back and dance again until three or four in the morning. . . . It was supposed to end . . . about one or two but then the orchestra would have a little hat or something out on the front of the stage . . . people would put money in the hat and that would be their payment for overtime," recalls Pat Goodwin.

"The dance floor itself had springs underneath it . . . when you danced, your legs would never ache because there's always give to the floor and you get a bunch of people doing the same beat . . . you never had tired legs the next day," adds Ed Goodwin.

Katie Conway Bell Adair mentioned the dances as well. "Different people would play . . . everybody played the violin and that sort of thing . . . and some people would play on the piano and some people from Bridgeport. . . . Slick Bryan had an orchestra he gathered together, you've heard of Slick Bryan in Bridgeport? Well, he owned nearly all of Bridgeport; his dad did, they had the dance hall in Bridgeport. We'd go to Bridgeport one weekend, then to Bodie the next when we were going to high school. . . . He played the saxophone. . . . They had real good music. . . . [W]e wouldn't get home until eight or nine o'clock in the morning. . . . My mother would feed everybody that came home with us breakfast. . . . I had a real nice childhood."

Ed and Pat (Dolan) Goodwin

Pat and Ed Goodwin are both Bodieites, although of varying degrees. Pat Dolan Goodwin was born in Bodie and spent summers there while she was growing up, clear through college. She spent her third- and seventh-grade school years in Bodie as well, living with her grandmother Alice McDonnel, who served as Bodie's postmaster (postmistress). The post office was in the front of her house, next to the bank. Only the bank vault remains in that area today.

The McDonnel House is named for her grandparents, who were the last people to live there. Pat's sister is Marjorie Dolan Bell Voss. Their granduncle was Sheriff James Dolan, who was killed in a shoot-out near Mono Lake. A monument stands there today to honor his memory.

Ed Goodwin's mother was May Cody, Ella Cody Cain's sister. Ella Cody married David Victor (or "D. V.") Cain, J. S. Cain's son. Ed never lived in Bodie; however, he was a frequent visitor and remembers it well.

Pat recalls that as children, they were not told to keep away from much, but "they all told you to stay out [of the mine area] because there was cyanide . . . opposite of where the Grays lived. . . . I suppose there was probably cyanide dust there. We never played in it."

Pat's uncle Frank once took her and a friend down into a mine shaft. "My uncle was working in the mine at the time. . . . [Y]ou just sat on a platform—they called it a 'cage,' but it wasn't a cage. It was a platform and I think there might have been something . . . one of those braided rope wires that went around it so you wouldn't fall off . . . and then you went down. I was afraid that they were going to blast while I was down there. Of course they weren't. My uncle would never take me down if they were going to blast; they wouldn't have let him in the first place. But we . . . got off [at] about the second level down and walked in. It was a pretty big area to walk in. I didn't like it at all."

When Pat was a girl, she and her friend Lucille liked to play in the old hotel across the street from her grandmother's house. "It was empty and we used to play there and . . . take buckets of water over . . . we cleaned it all up. It belonged to old J. S. Cain. He did not care; he let us go over there and play and he would even let us play with all the ledgers. We used to . . . sign new people in. We would pretend we were renting rooms. When we got it all clean, he rented it!"

When Ed was a child, he and a pal played in the defunct casino one day and toyed with the roulette wheel. "I used to throw the ball in there and spin it. We looked underneath it and . . . a dealer could make a pin come out of one of the holes. Yeah, it was crooked! . . . I remember going to the under

ROSEKLIP

In 1935, Jack Rosekrans paired up with Henry Klipstein to form the Roseklip Mining Company. They leased mines from the J. S. Cain Company and went to work reprocessing old tailings using the cyanide process. They worked in Bodie with varying degrees of success, from 1935 until 1942. They mined 55,000 tons of dump material that they retrieved with "mechanical shovels." Old-timers say the Roseklip mill ran around the clock, because that was the only way they could make a profit.

Gunnar "Pete" Peterson, assayer for the Roseklip operation, remembered his days in Bodie well. "We lived right next to the Standard mill. My wife was the postmistress. . . . [O]ur assay office was on top of the hill where the mill burned down . . . in '46."

Although Roseklip leased the land on the hill from the J. S. Cain Company, it owned its own mill. "[I]t was a ball mill . . . it was a cyanide plant. We were treating about 500 tons a day of 'dump ore' . . . from the Bulwer

taker's place and smelling the bottles—opening them up and [the] terrible, horrible smell of the stuff—formaldehyde, I guess. It almost made us sick!"

Growing up, Ed Goodwin heard all kinds of things about his Aunt Ella Cain, including a local myth that she had a "mummified head" upstairs ("I don't remember that," cautions Pat). His aunt was indisputably famous for her Indian basket collection, however. "She had them made by the local Indians . . . she bought them . . . and then she would show them in the Sacramento State Fair," says Ed. "She paid them for the baskets. She won a lot of ribbons at the fair for her collection." In the late 1900s, Ella Cain's collection was put on permanent display in the Mono County Museum in Bridgeport.

Pat recalls, "[O]ne time [Ella and Vic] took a trip down to Aurora and when they got back somebody had been in their house. They assumed that they got up and left . . . with their food still on the table. Well, they had, they had taken a drive after dinner! They hadn't left [for good]. They didn't lock the house."

Ed adds, "Like the Grays. They'd be up there all summer. They didn't want to stay up there in the winter. They would just close the house up like anybody would . . . a summer home. You'd close your house up, prepare it for winter and leave part of the stuff in there." Both Pat and Ed say the idea that people left things in Bodie when they moved away is unlikely. Most people took their possessions with them, and the rest is just a popular misconception.

tunnel." This was far more ore than the Standard Mill handled in its day, according to Peterson: "[M]aybe the stamp mill would go 50 tons a day. . . . They [Roseklip] also had a smelter up on the hill. They used the Merrill-Krull process to recover the gold from the cyanide solutions."

The ore they processed brought between $2.50 and $5 a ton. Five dollars a ton was enough to make a profit, "because wages weren't very high in those days." This is quite a contrast to the days of the Burgess vein in June 1878, when ore assayed at $3,000 per ton.

"We had between 20 and 25 [workers] fluctuating, working at the mill and at the mine, operating trucks and a power shovel. At night, for two shifts, we'd only have two; one on the ball mill, one on the mill." The third shift, the day shift, miners worked as well as mill workers.

There were no other mill operations in Bodie then, according to Peterson. Henry Klipstein lived with his family in a house on the hill in Bodie. Jack Rosekrans leased and fixed up the old railroad depot building at the top of

Green Street. He was famous for entertaining guests lavishly. He even hired a cook and a housekeeper for his summer guest schedule. His guests would explore Bodie and the old mine shafts, hunt for sage hen, fish in the storage pond near the house (which was stocked with trout), or shoot skeet. The storage pond was there as an emergency water supply in case there was a fire on the hill.[7]

END OF THE BODIE MINING DISTRICT

World War II brought a halt to mining and processing activity in Bodie. Most of the nation's available resources, human and otherwise, were diverted to help the war effort. There was also discussion about the possibility that the lights used in the night could attract enemy aircraft. The final blow was President Roosevelt's decree shutting down all mining work that did not directly aid the war effort. "That L208 closed us down," Gunnar Peterson recalled.[8] "November '42 is when the post office closed[9] and that's when we moved out. . . . There were a few people still here; there was a watchman who worked for Cain."

After World War II ended, most Bodieites came back for the summers, but they spent the rest of the year working elsewhere.

In 1945, the J. S. Cain Company (J. S. himself had died in 1938) leased some mines to Sierra Mines Inc. to confirm that there was an abundance of "marginal ore" in the Standard Hill area. An accidental fire destroyed the Syndicate mill and cyanide plant in April 1946.

THE GHOST TOWN OF BODIE

> *It is beyond understanding what people will do.*
>
> —*EMIL BILLEB* *([1968] 1986)*

After the mines closed down, the Cain family hired caretakers to look after the town, but they could not be in all places at all times. There were three caretakers in Bodie for several years, living there year-round. Astoundingly, the Goodwins say, two of them got angry over something and would not speak to each other! Not even in winter!

Spence Gregory, one of the caretakers, was from the Bodie Ranch, on the road between Bodie and Aurora. He was Dorothy Joseph's uncle, and she spoke of him fondly. "He spent the summers in Bodie . . . on Green Street. The Bodie Ranch had deteriorated so, when he was working in the museum and so, he was living in there [in the Gregory house]. . . . They redid it these last two or three years, which I was very grateful for because it needed an awful lot of rehabilitation. . . .

"He was an engineer . . . he was [paid by the Cains] a very small amount, but he did engineering too. . . . [I]t wasn't a going proposition. He used to work with my uncle as an assayer. . . . He would pick up things in the mining field."

After Bodie became the dominion of hired caretakers, Bob Bell remained a frequent visitor. Bob, like other Bodieites, recalls the caretakers were an odd bunch. Apparently, Spence Gregory was not an overly friendly fellow. When Bob Bell visited one of the other caretakers, he asked after Spence. "The fellow told me, '[W]ell, I think he's O.K. I look over every now and then and I see smoke coming out of his chimney.' I said, '[D]on't you go over and visit him?' 'Well, I went over to see how he was, but he played "freeze out" with me,' which meant that he stood outside and didn't invite him in, until they were done talking!"

Bob says there once had been another caretaker, a man who was rumored to enjoy a nip now and then. He would drive his car with a glass of wine wedged between his foot and the floor. This fellow was fond of carrying a shotgun with him for enforcement when he found intruders in the town. One night when Bob was with him, a couple of fellows came into Bodie, and they heard the men shoot what turned out to be a snowshoe hare. The illegal hunters headed quickly up the hill, but the road was chained off. They

Dorothy Joseph

Dorothy Joseph was born in a house about halfway between Bodie and Aurora: "[I]t was a stagecoach stop . . . called the Bodie Ranch." The ranch was owned by her grandparents the Gregorys. Spence Gregory, one of the last Bodie caretakers, was Dorothy's uncle. In October 1987, Bodie SHP volunteer Diana Mapstead interviewed Dorothy Joseph.

Dorothy Joseph didn't live in Bodie as a girl, but she did have memories of the Bodie Ranch and the lawless atmosphere of the area. "When my grand-father Gregory would bring the cattle in, somebody was always sitting on the corral with a gun to protect him. My grandmother always slept with a Colt revolver on her bedstead, because the ranch sits back by the hills and all these robbers and what come over the hill . . . it was danger all the time.

"I never really lived up there [in Bodie], but Mother was raised there. Her name was Idelle Gregory . . . she was the second child. . . . My mother and her sister, Christine, at one stage of their existence, put out a little newspaper: *The*

Miners' Union. I have a copy of it. . . . Actually, there isn't any news in it, it was copied stuff . . . it existed and they sold it." *

Dorothy Joseph said her mother's generation in Bodie faced a different seasonal school schedule. "The weather changed, the whole cycle of the weather. They were lucky if they could get two or three months of school in a year. And then my grandfather was a very strong disciplinarian and they would study at home."

In Ms. Joseph's copies of *The Miners' Union,* park volunteer Mapstead found a description of a ball attended by one of Ms. Joseph's ancestors. Volunteer Mapstead read out, "1873 journal from Reno of the May IOOF ball, attended by 100 couples. Mrs. Spence Gregory is described as a demi-blonde, dark eyes, auburn hair wearing green silk with puffs of illusion . . . secured with pink flowers on the side, cut surplus waist with illusion bertha, hair a la grecque with flowing curls with white and green sprays in same."

However, because Spence Gregory was known as a lifelong bachelor, Ms. Joseph had to think a moment about who this Mrs. Spence Gregory could have been. "My uncle Spence was named for his uncle . . . that must've been where the other Spence Gregory came from."

She remembered her Uncle Spence very fondly. "He never married, so he lived there. . . . [H]e took over the care of anything . . . along in the '30s and what have you. . . . He and my husband became close friends, they were both World War I veterans." Spence Gregory's portrait hangs in the Bodie museum.

---

* Interviewer Mapstead noted the paper's masthead: "I. D. Gregory, proprietor, C. H. Gregory, Editor. So no one would know they were women. . . ." "No," agreed Ms. Joseph.

didn't see the chain in the dark, ran into it, and got stuck. The caretaker and Bob Bell quickly headed out in the caretaker's Model T with the shotgun, and sped up after them. As they reached them, the caretaker ran out with his shotgun and poked at the window with it, trying to get at them. The hunter came out, and when the caretaker asked them, "What're you doing here?" he said, "We're lost." "Well you just go back down there and you won't be lost anymore," he said. They agreed and hightailed it.

Another time, Bob and his family were up on the hill and saw some folks up near the Standard Mill. The caretaker roared up there in his Model T, leaped out with his shotgun, and accidentally fired the gun, blowing a hole in the ground. The people got scared and said, "Hey, you wouldn't shoot a fellow, would you?" He hollered, "Just try me!"

This fellow may have been a bit over the top, but his methods certainly discouraged vandals and thieves!

GRAVE ROBBERS AND VANDALS

The unfortunate modern myth of Bodie's abandonment led to many instances of burglary and vandalism. At one point, someone even stole the sizable bell off the firehouse roof (it was later traced, found, and returned). The cemetery was practically destroyed by vandals. To the astonishment of the caretakers, on one occasion people were found there with metal detectors, looking for guns buried with their owners. They were apparently willing to become grave robbers.

Ed Goodwin says, "People used to come up and loot Bodie. . . . They stole a lot . . . they even took headstones away from the graveyard. . . . [I]magine people were looking at this old abandoned town that still had furniture and they just went in there and raped the place."

Gordon Bell witnessed some attempted vandalism firsthand one day. "Emil Billeb and I were up there in Bodie. . . . He came down, . . . some guy had broken down the door into the Hydro building. Emil said, 'What're you doing in here?' The guy says, 'Well, there's no signs out there that say "Keep Out." ' Emil says, 'Where do you live?' The guy says, 'Down in southern California.' Emil says, 'You got a sign that says "Keep out"?' The guy says, 'No.' Emil says, 'I'm going down and kick *your* door in. . . . You fix that door, otherwise you're going to jail.' " The man fixed the door.

Still, as Gordon sadly points out, many vandals and thieves have made off with bits of Bodie, even bits anyone should recognize as extremely personal. "That cemetery used to be loaded with all kinds of headstones; wooden ones as well as the other ones. They've all disappeared."

Bodie will always be "up home" to many people still alive who visit on a frequent basis. And Bodie's dusty cemetery up on the hill is the permanent final home to many others. Visitors may enjoy looking at Lottie Johl's triumph of final placement, or the hauntingly dear angel for Evelyn, but there are many

Many people's loved ones rest forever in Bodie.

This sad angel watching over little Evelyn's grave attracts many visitors each year.

The Up-To-Date manufacturing company crafted many of the grave-site fences in the Bodie cemetery.

other people who come to pay their respects for fairly recently departed loved ones. It is not an abandoned graveyard full of forgotten names. Almost all the Bodieites interviewed for this book have loved ones laid to rest in the Bodie cemetery. Some of the Bodieites laid to rest in the cemetery are honored in unusual ways. There is at least one family that remembers its dear departed patriarch by pouring a bottle of beer over his grave. He always liked it much better than flowers.

8

In the days of old, in the days of gold,
how oftimes I repine,
for the days of old when we dug up the gold,
in the days of forty nine.

"Days of '49"—a Gold Rush-era song
(Arlen, Batt, Benson, and Kester 1995)

Bodie State Historic Park

In 1958, the State of California, at the behest of the Cain family, took a long, hard look at Bodie's potential as a State Historic Park (SHP). The state officials reviewed records, inventoried what was left in the town, and considered several possibilities. Finally, when they decided to indeed move forward, it was with a new and novel approach: arrested decay. No fixing up, just preventing Bodie's further disintegration.

Emil Billeb, who first came to Bodie in the spring of 1908 to work for the Bodie and Benton Railway and Commercial Company, served as representative of the J. S. Cain Company in negotiations with the State of California. He was married to J. S. Cain's daughter, Dolly. Billeb and Dolly lived in Bodie until 1920, when they moved to San Francisco. Billeb's negotiations with the state resulted in the creation of Bodie SHP.

The Cain family sold all their property in the Bodie townsite to the State of California, but held on to their property on the bluff. The next step for the state was to obtain the remaining properties within the Bodie townsite, which were owned by several different families.

It seems clear now that the only way Bodie could have remained at all intact would be to follow the path it has, to become a treasured historical park. At the same time, for the Bodieites who were suddenly faced with an offer of a few hundred dollars from a state agency for their family property, full of personal history and sentiment, it seemed a rude and faceless shock. This was an agency coming at them, an agency that could give them a small amount of money or could condemn that property and force them to sell it. They had little choice. The state arranged to purchase the remaining houses and land, and created Bodie State Historic Park.

The Bells were paid for their houses in Bodie, but like most of the Bodieites, they had a hard time accepting money as appropriate compensation for the loss of their hometown and homes.

Marjorie Dolan Bell Voss was offered a few hundred dollars for her house. She sued for $2,500, because at that time, similar "lots in Bridgeport . . . were going for that and I would rather live in Bodie than Bridgeport." She represented herself in the proceedings. She won $500, which was more than many others got for their family properties.

The Grays tried a different approach. "When the state bought the park, we asked for occupancy rights, to continue to visit in the summer and keep our house up. The state refused. I sued and lost," says Fern Gray Tracy. In retrospect, that might have been an interesting path for everyone: With the occasional resident in for a stay, the town would have been even more intriguing.

"When we had to move our things, we moved them to the houses on the hill [houses owned by the Cains]," says Fern. For the time being, the hill

A watercolor portrait of the Gray family home before Bodie became a state park. Courtesy the Gray-Tracy Collection

houses remained in the Cain family. "We kept our stove and beds there for when we came deer hunting each year. Then, of course, one year all our stuff was ripped off, the top of the stove gone, the rest dismantled there in a heap. We were angry."

Mrs. Lauretta Gray hadn't seen her house since it was bought by the state in the early 1960s. "Our house has disintegrated since then," says her daughter Fern. The family remains unsettled about the changes to their cherished family home in Bodie, and it seems unreasonable to them to pay an entrance fee when they come visit. And who can blame them? They grew up here, had homes here. Few of us have to pay to visit our hometowns.

Happily, the park staff came to recognize this. The last time Fern Gray Tracy visited her childhood home, the rangers welcomed her. They "opened up the house and walked all around town with me to get the proper story," she says. The Gray family home is not what it was when they lived there, but it has been stabilized to prevent further decay.

At the turn of the twenty-first century, by and large, the Bodieites were pleased to see that their town was being preserved as a park, although many say it is strange to see your hometown presented as a tourist attraction. Understandably, it doesn't seem quite right to them; the town isn't exactly as they remember it, and remember it they do. It helps that members of the park staff recognize that the Bodieites are a crucial link to the park's history.

When the first park rangers arrived in 1962, there were still a few hardy souls living in the town along with the caretakers. Some of them ran the museum Ella Cain had set up in the Miners' Union Hall.[1] The rangers' first months were spent trying to keep generators going so they would have power, fixing various leaks, and keeping an eye on visitors. As they progressed, they learned how to best manage the town to preserve it for the future. Former Bodieites proved to be a valuable resource. Several of them assisted the rangers, showing them where things were and what preparations were needed to keep things running smoothly. Bob Bell became so invaluable to them, they repeatedly hired him as a park aide. Through trial and error, successful management techniques emerged. Limiting vehicles to an entrance road and parking lot was the first major improvement. This not only made it safer to stroll through Bodie, but also dramatically reduced the amount of thievery.

Bodie SHP was officially dedicated on 12 September 1964. Slowly, conditions for the on-site staff improved, although there were no phones and electrical power was an on-again, off-again kind of thing. And the fragility of the structures was apparent, as this 9 May 1966 ranger's report illustrates:

Bodie in pop culture, 1995. Courtesy Bill Griffith

Yesterday we witnessed what one would expect to see no oftener than once in a lifetime: The destruction of a house on Green St. by whirlwind. This house is the fifth one east of Main St. on the south side of Green. The winter winds removed the roof and left the walls standing at various angles. By a one in a million chances [*sic*] we stepped out of our residence just in time to hear the rending of wood and the screech of nails being drawn from the wood and watched the whirlwind play havoc with the remaining parts of the structure. What was left partially standing has been braced and nailed to reduce hazard.

As time progressed, managing the park has become no less dramatic or time-consuming, but the tasks are more familiar and established. The state has learned how to handle its experimental baby, and it has prospered. Then, as the price of gold rose to new heights, interested eyes began to examine Bodie once more.

THE ARGONAUTS RETURN

> *The mine is always bigger than the gem.*   —*Sufi proverb*

Because Bodie owes her entire existence to the frenzied pursuit of gold, it should not be surprising to learn that there are contemporary Argonauts interested in getting at any gold that might be left.

The gold-mining industry is fueled by consumers' demand for gold. In the United States, in the 1990s, 22 percent of the gold produced was used in the electronics industry, and 71 percent was used in the jewelry and arts industries. Just a smidgen was used in dentistry, where we perhaps benefit the most directly from it. On the global scale, 6 percent went to electronics, 85 percent toward jewelry, and 9 percent toward "other." The pressure to dig new open-pit mines and cyanide ponds really does come from the demand for more cuff links and bracelets. It can be hard to realize that the new jewelry on sale cheap at a department store is what lies behind the carved-away mountain landscape that is so startling to behold. Yet it is so.

In the late 1960s and early 1970s, a few different companies leased properties, conducted studies, and took ore samples on the surface and in easily accessible mines. None of these explorations resulted in active mining operations. In 1972, the first core drilling–style prospecting began.

In 1988, Galactic Resources Ltd. of Canada formed Bodie Consolidated Mining Company and began a feasibility study, allegedly with an eye to establishing at least one open-pit mine and a cyanide heap-leach operation on the eastern side of Bodie Bluff. How much buried treasure did Galactic expect to find? The number has fluctuated some. Today, former employees agree they expected to remove between 1 and 1.25 million ounces of gold. This is an impressive figure when you consider that during the entire Bodie heyday, at most 1 million ounces of gold were removed from the "bonanza" area.

Various reports of the time say that Galactic planned to process about 230,000 ounces of gold each year for somewhere between 10 and 20 years.[2] At $390 an ounce, that's $89.7 million a year, almost a third of California's 1988 gold production of $320 million.

Those are big dreams, and the gold business is fueled by them. Just to get to their feasibility study, Galactic had to invest quite a lot of money. Depending on which report you read, the company paid the Homestake Mining Company (of San Francisco) between $31 million and $39 million for the mining rights to 550 acres of private lands (owned mostly by the J. S. Cain Company).[3]

The battle lines were soon drawn in Mono County, with the predictable camps of environmentalists ("Just say NOPE: No Open Pits Ever") and mining enthusiasts ("[Bodie] was a rambunctious mining town, that's what Bodie is. That's what Bodie should be") quickly established. Many others remained uncertain about the situation.

The objections to mining operations in Bodie (especially those using the cyanide heap-leach method) centered mostly on three issues: ongoing disturbance of "the Bodie experience," destruction of the buildings in the state park from nearby blasting, and destruction of local wildlife from ingesting cyanide solution from cyanide ponds.

The Bodie experience is the hardest to quantify on the Environmental Protection Agency forms. Bodie's haunting silence is an integral part of its lessons and its testimony. Bodie is best when, for just a moment, you're not sure what year it is or, perhaps, just who you are as you stand where miners and others lived and died, years before.

The ongoing drone of modern ore trucks and drilling rigs (let alone the occasional blasting operation) would greatly disrupt this experience. The ambiance of Rhyolite, a ghost town bordering Death Valley, is very different from that of Bodie because of this modern presence. Rhyolite, although interesting, is the discarded remains of a town just over the hill from an operating gold mine.

Blasting operations destroy what is in their path: That's the whole point. Luckily, the force can be directed to a degree, which is why it's so useful in mining and road building. However, there are several reports of modern houses with cement slab foundations (as well as old structures in Rhyolite) being damaged by mining blasts several hundred feet, to a half mile and even one mile, away. The hastily built wooden structures of Bodie wouldn't stand a chance under such repeated assault.

Wild animals die when they drink cyanide. Unfortunately, they cannot be easily deterred when they think they've stumbled upon a magnificent pond in the middle of an extremely arid region such as Bodie. Many mining operations have tried all sorts of tricks to shoo away birds, mammals, and others from their cyanide ponds, but between 1984 and 1995, more than 13,000 animals were killed from these ponds and/or other chemical poisoning in Nevada.[4] Adding to the difficulty, in an area as sparsely populated as Bodie, what seems like a relatively small number of animals can constitute a large portion of the existing ecosystem.

ALL'S WELL THAT ENDS WELL

The cleanup costs and fines from Galactic's disastrous operation in Summitville, Colorado—an operation of the same type it was considering in Bodie—bankrupted the company, more or less calling a halt to its aspirations for Bodie.[5] Galactic left an enormous mess in Colorado for the EPA to clean up, costing thirty-three thousand taxpayer dollars a day, and closed up shop, selling off all its assets save one: its Bodie holdings.

After Galactic declared bankruptcy, its employees scattered. Bodie Consolidated Mining Company eventually put its Bodie holdings up for sale. Various personnel from the State of California and the U.S. Bureau of Land Management (BLM) scrambled for a plan and funds.

In 1994, while Bodie SHP's future lay in the balance with its neighboring land and mineral rights on the block, the U.S. Congress passed the California Desert Protection Act of 1994 (the Desert Act). The Desert Act included the Bodie Protection Act, which asked the federal government to review all the existing mining claims in the area and stop any new ones from being issued. Unless they could find a compelling reason to risk the Bodie townsite's very

Home Comfort stove

existence to allow gold mining in the area, they were to declare the claims not valid. Although the Desert Act helped Bodie, it by no means ensured Bodie's continued protection. As long as there was significant money to be made from gold in the Bodie Mining District, there was cause for concern about the ghost town of Bodie.

As Galactic's bankruptcy put pressure on its board to sell its Bodie interests quickly, personnel from the State of California and the BLM sought sufficient funding to purchase the land and its mineral rights, to secure it in public hands for the foreseeable future. In 1993 a tentative deal was reached by which the California Department of Parks and Recreation (CDPR) would make the purchase for $5 million. CDPR was able to secure $3 million in grants, for a down payment, but the remaining $2 million was hard to find. Galactic wanted and needed the money immediately: CDPR stood to lose its chance to make the purchase if it didn't act quickly.

Enter the park's savior in the form of Harriet Burgess and her American Land Conservancy (ALC) organization. The ALC provided the remaining $2 million, closing the deal with Galactic in 1997. In October 1998, the BLM finished reimbursing the ALC with federal appropriations funding, placing the land and its mineral rights fully into public hands. This acquisition increased the size of Bodie SHP from 495 to 1,015 acres.

After one last decade of gold-based excitement in Bodie, and efforts to both produce and ward off another mining boom time, the resolution of the situation brought sudden silence once again to the high-desert townsite.

AND NOW . . .

Bodie stands silent and empty, garbage dumps full of rusting cans and broken bottles, old bedsprings and cars, cyanide ponds and heaps of tailings—stained ochre piles like enormous gopher mounds, pockmarking the landscape. The bad men who killed and were killed are gone, the vigilantes are gone; all are gone from the dry, windy bowl. And slowly, slowly, Nature reclaims what is hers. Patiently, the buildings are decomposing. Lack of budget funds keeps the state from saving every one. "Just the significant ones, that's all we can afford." And so Bodie stands also as a monument to human civilization and its values: a force that can take breathtaking desert landscape and turn it into enormous monuments of refuse. And yet Nature works even with that and slowly, painfully, works to reclaim what was only borrowed.

And it came to pass on the third day in the morning,
that there were thunders and lightnings,
and a thick cloud upon the mount . . .

. . . and the smoke thereof ascended as the smoke
    of a furnace
and the whole mount quaked greatly.
                        —*Exodus 19:16 and 19:18*

Appendix 1

Once Upon an Ancient Time . . .

Bodie's violent geologic history set the stage for its lively human settlement, because it was the ancient volcanic activities that put the gold there. To determine the approximate time line of these geologic events puts us in the hands of geologists, who at any given moment have the best theory possible but cannot promise that it will stay put.

According to the current theories of plate tectonics and geologic history, the entire region that is California is a relative newcomer to the globe, having come together in bits and pieces via multitudes of earthquakes large and small. The Bodie area (and the region extending a ways south as well) was once home to steaming and burbling volcanoes that spewed forth much of what we park next to and on top of today. Mammoth Mountain, a volcanic vent not far south of Bodie, is still grumbling today; the region is not entirely quiet yet.

The rocks of any area offer the local geologic résumé. Their structure indicates how they were formed and therefore what happened in the region. Their age indicates when it happened. Bodie rocks are virtually all of volcanic origin. Various igneous and/or metamorphic rocks were formed at different times, and because the land has curved and buckled here and there, you can find the very old alongside the relatively new.

West and northwest of the Bodie Mining District lie metamorphic and granitic rocks that formed during the Cretaceous period (70 to 132 million years ago). And at Masonic Mountain, just twelve miles northwest of Bodie, you can find older metamorphic rocks from the Jurassic period (132 to 180 million years ago). In the Conway Summit area, about twelve miles south-west of Bodie, you can find still older metamorphic rocks from the Paleozoic era (275 to 600 million years ago).

The geologic makeup of the Bodie Mining District itself—"inter-layered lava flows and tuff breccia" (Chesterman, Chapman, and Gray 1986)—is the hardened remnants of magma that came all the way out of the volcanic vent to the earth's surface. Tuff breccia is a combination: *Tuff* includes the hardened rocks from ash falls, ash flows, and other types of material ejected from the volcano; *breccia* is a goulash of sand, clay, or lime mixed with sharp bits of fragmented rock.

The igneous rocks of the Bodie Hills date from the later Tertiary period (a mere 12 to 70 million years ago). The oldest of these rocks, in the northeastern part of the hills, are between 11 and 29 million years old. Here and there over these is a layer of "ash flow tuff" (Chesterman, Chapman, and Gray 1986) that is, on the average, about 9 million years old: a mere geologic toddler. The Bodie Hills volcanic rocks may have come from Bodie Bluff, Queen Bee Hill, and Sugarloaf, all of which geologists consider to be volcanic vents that had several very active periods ranging from relatively quiet times of lava flows to explosive episodes that left the tuffs and tuff breccia. There are other volcanic vents nearby, such as Bodie Mountain, Potato Peak, Mount Biedeman, and others, but Bodie Bluff, Queen Bee Hill, and Sugarloaf are the only known vents within the Bodie Mining District.

DOMING

Geologists believe that all the Bodie-area vents were connected to the same magma chamber under the earth's surface, although that doesn't mean the vents were all active at the same time.

At some point, hot magma from below intruded into the lava flow and other rock layers above. The magma, which was less dense but hotter than the surrounding rocks, rose toward the surface through the other rocks, bending and lifting up (called *doming*)[1] the existing layers. This is what gives the area its characteristic appearance of "a series of hills amid flat lying layers" (Silberman 1997). The doming produced fractures within the layers on top that opened up the rock to alteration and mineralization, which resulted in the deposition of gold and silver ores.

---

1. A straightforward but unappealing analogy is what occurs to people in adolescence when the skin is prone to "eruptions" or pimples. The buildup of substances and distortion of the skin that go on are similar to geologic doming of Earth's surface. The description of the analogy can safely stop there.

The Bodie-area volcanic activity drew to a close between 8.8 and 9.4 million years ago, when "plugs" of igneous rock formed and filled the dying volcanic vents. Each vent was sealed with a similar plug, which halted the spewing of magma from below. The largest plug in the district is under Bodie Bluff and Standard Hill. The Bodie Bluff–Standard Hill and the Queen Bee Hill plugs appeared about 9 million years ago (give or take a few hundred thousand years), earlier than those at Sugarloaf and near the Red Cloud Mine, which are about 8.5 million years old.

Although the vents may not have all been active at exactly the same time, their age range indicates that their individual plugs all appeared within a relatively short time span, geologically speaking, of "on the order of one to one and a half million years" (Silberman 1997).

THE ELEPHANT ARRIVES

Shortly after the plugs appeared and the volcanic activity stopped, the elephant of gold followed.[2] How it got there is pretty much agreed upon by geologists, who point to hydrothermal fluids as the benefactors.

Hydrothermal fluids alter rock (hydrothermal *alteration*) when fluid (usually mostly water with dissolved silica-based mixtures), which has been superheated by magma, shoots up through the cracks and fissures in the rock. This very hot fluid, which is under high pressure, contains a lot of dissolved minerals. As the fluid progresses upward through the cracks, it cools, leaving metal and mineral deposits along the cracks and fissures. This process is called *mineralization.*

How hot does the fluid have to be? Water begins to boil at 100° Celsius (212° Fahrenheit) under normal conditions at sea level (at 8,500 feet, at the top of Bodie's elevation, water boils at 92°C/198°F). Silica-based liquids boil at even higher temperatures. However, the intense pressure keeps the fluid liquid at very high temperatures, instead of allowing it to vaporize. One study of "trapped bubbles and gases in minerals associated with the gold and silver at Bodie indicated temperatures of about 250°C" (Silberman 1997).

---

2. The expression "seeing the elephant" was popularly used to describe people going to the goldfields. The expression is believed to have originated before gold rush times, from traveling circuses. The defining tale is of a farmer who loaded up his wagon and headed to town for market and to see the visiting circus there. On the way he saw the circus parade, headed by the elephant. The sight spooked his horse, which tipped the cart and ruined his produce. " 'I don't give a hang,' the farmer said, 'for I have seen the elephant' " (Levy 1992).

When some of the magma from the plug at Bodie Bluff moved to form the Sugarloaf and Red Cloud plugs (from their shared magma chamber), faults developed in the Bodie Bluff–Standard Hill area. Two of these faults are the Moyle Footwall and Standard Vein faults.

Movement along the various faults in the Bodie Mining District opened up cracks that could then be filled with gold, and associated minerals such as quartz, via the aforementioned *mineralization*. When faulting, or movement along existing faults, occurred after the gold veins had formed, many veins were cut into segments. In some places where existing veins were segmented this way, new mineralization occurred along those cracks and breaks, form-ing new veins. In general, the intersection areas where new veins met old were richer in gold than other areas. Miners referred to these areas as "enrichers."

SEARCHING FOR GOLD

When work in the Bodie Mining District first began, most of the metal ore was found within 500 feet of the earth's surface. One vein, the Fortuna, was an exception and was mined down to about 600 feet below the surface. When mineralization first took place, there was, in some areas, an additional several hundred feet of rock between the surface and the deposits, rock that has since eroded away.

The layout of the Bodie Mining District ore bodies made the miners' work fairly difficult. Bodie has a lot of *fissure veins* that are laid out rather like the uppermost branches of a tree: Many veins expand out, growing increasingly thin the farther out they go. These are difficult to trace and get at for gold extraction using the underground mining techniques of the 1880s. The work was labor intensive, and the mines themselves required a lot of supportive framing.

This type of ore body structure is more effectively and economically mined using the bulk-style techniques common today (described in appendix 2, "*De Re Metallica* Redux"). In comparison, an ore body like that of the Comstock in Virginia City, Nevada, which was large and straightforward for extraction, was more easily mined using the underground techniques of a century ago, although more modern open-pit methods have been used there as well.

In geologic terms, the gold in the Bodie Mining District did not age long before humans began to harvest it. Tests performed on material from various areas throughout the district reveal the gold to be between 7 and 8 million years old.

De Re Metallica Redux

The original *De Re Metallica* was written in 1556, by a Saxon named Georgius Agricola. This work describes state-of-the-art mining procedures and tools of the time. It is a standard mining reference text, still used by many modern mining scholars. This appendix is an abbreviated explanation of some of the methods and tools used at one time or another in Bodie. Also included is a brief overview of the cyanide heap-leach process, which was proposed for use in Bodie in the 1990s by the Bodie Consolidated Mining Company. Although this is by no means an exhaustive study, it gives the inexperienced reader some idea of the tools and methods found in Bodie over the years.

IN THE BEGINNING: ARRASTRAS AND ORE ROCKERS

W. S. Bodie and E. S. Taylor had no stamp mills to pound up large quantities of rock; they were limited to equipment that they could haul with animal help, which meant slower processing. Prospectors generally used arrastras (sometimes called *arrastres*) and ore rockers.

Arrastras have been around a very long time. Mexican miners introduced them to the California mining world. They are still in use in some places today. The technology dates at least as far back as 1556, because it is described in the original *De Re Metallica*. An arrastra works on principles somewhat similar to a pepper mill. The flat, circular grinding surface is sunken in the middle, raised on the inside and outside edges. The surface is made of rocks jammed in together. Heavy grinding stones, called *drags*, are chained to wooden arms that drag them around the grinding surface. The drags are set so no ore can escape being pulverized. The arrastra can be powered by water, people, or animals. (For a photograph of a typical arrastra, please see chapter 2.)

The miner put the ore in a little at a time, adding just enough water to make a pulp the consistency of thin mortar. If the pulp was too thin, the crushing took longer. Five to seven hours of grinding later, the pulp was fine enough to add quicksilver (mercury). After adding the mercury, the miner ground the mixture for two to three hours longer. The gold dissolved in the mercury, leaving the rest of the rock and sand behind. This process is called *amalgamation*. When enough gold was taken up by the mercury, the amalgam became partially solidified and settled into the crevices of the rock-grinding surface. The miner then picked the gold out of the crevices using picks, scrapers, and spoons.

Although most arrastras used water, an interesting water-free design appeared in Bodie in 1884. According to the *Manufacturer and Builder, a Practical Journal of Industrial Progress,*

Out in the Bodie mining district, California, they have a peculiar motor in use. It is called an arastra [*sic*], and consists of an overshot wheel operated by sand instead of water. A wind-mill runs a belt containing buckets, which carry the sand up to a big tank, just as grain elevators carry wheat in a flouring mill. A stream of sand is let out upon the overshot wheel and it revolves just as it would under the weight of a stream of water. The arastras move steadily at their work. When there is much wind, sand is stored up for use when calm weather prevails, so the arastras are never idle. It is perhaps needless to say that the sand is used because water is scarce. The arastra is an invention of a miner named Townsend.

Ore rockers were short troughs set on cradle rockers tilted to one side. One miner loaded the graveled ore onto a top screen that sifted it onto the rocker as water was poured through. Another miner rocked the rocker, nudging the ore through. As it passed over the *riffles*,[1] the gold, which was heavier than sand, got caught in the riffles and shifted to one side of the rocker. Ore rockers and most arrastras (with the mentioned exception) required a fair amount of water to work properly, a difficult proposition at times in the Bodie area.

---

1. Riffles are slats across the bottom of the rocker, designed to catch gold particles as the water passes through.

The Bodie gold-processing business took an upswing with the construction of the first *stamp mill*. Stamp mills could process ore on a much larger scale than could arrastras and rockers. The stamp mill had several levels, and its design allowed gravity to aid the whole process. The ore was dumped into a *grizzly*, an iron grating similar to a cattle guard. Chunks of rock larger than 1.5 inches across could not pass through the grizzly and were directed into a crusher, which reduced them to the size of corn kernels. The smaller pieces passed through another grizzly and on to the stamp mill, where 900-pound iron stamps reduced the ore to a powder. The noise from the pounding stamps was intense. Mill workers kept beeswax in their ears to keep from going deaf. Those who did go deaf generally lost their job as a result.

After being crushed by the stamps, the powdered ore passed through a "40 mesh" screen, a screen with 40 openings per square inch. The powder was mixed with water, and the resulting slurry flowed over a copper-plated amalgamating table that had been painted with mercury to catch the gold. The runoff from the amalgamating table, which might contain some gold they wanted to save, passed onto a shaking table, which, as its name implies, shook, helping the heavier gold and silver to separate from the lighter gangue minerals (minerals with no commercial value). (For a photograph of amalgamating [stamp mill] tables, please chapter 3.)

The mill workers scraped the amalgam off the amalgamating tables and heated it in the retort room. Mercury vaporizes at a lower temperature than the metals do, so as the mercury-gold mixture heated, they caught the mercury vapor (and cooled it off, which returned it to liquid to be used again), leaving the gold.

The ore had both gold and silver in it, because both dissolve in mercury. The final product would be an alloy of gold and silver, and generally it was poured into bars still mixed. The gold and silver were separated out in the refining process in Carson City or San Francisco. Depending on the ratio of gold to silver in a particular bar, a "gold" bar could look more silver in color than gold.

BALL MILLS

The *ball mill*, which came into vogue somewhat later, was a variation on the theme of the stamp mill. Instead of pounding rock with heavy stamps, ball mills crushed the rock inside rotating cylinders that were filled with metal balls, rather like enormous rock tumblers. The *autogenous mill* used large rocks instead of metal balls in the cylinders to crush the ore. These rocks would eventually wear down to the same size as the ore and would be processed with the rest of the ore. The rocks would be replaced with other large rocks, saving the mining operation the cost of metal balls.

CYANIDE PROCESS

The cyanide process works under a similar principle as the mercury amalgamation, using cyanide as the extracting agent. The advantage of this process is that it is inexpensive and is used most effectively on low-grade ore. In Bodie the tailings piles from ore processed before the advent of cyanide were processed again, using the cyanide process, wringing the last bit of gold out.

The cyanide process became popular all over, but not with as much fanfare as the old days. "It was a very dull business, and it did little to stimulate moribund mining camps since it employed so few men. All it did was double the world's annual production of gold. There were, to be sure, big strikes and feverish boom camps after 1893—such as Goldfield and Tonopah . . . —but somehow things were not the same" (Young 1970). The cyanide process was used in later years in Bodie.

CYANIDE HEAP-LEACH METHOD

Today, we have a variation of the original cyanide process: *cyanide heap-leach technology,* which, partly because of its bulk-style, passive approach, makes processing low-grade ore particularly economically feasible. Using this technology, it's possible to retrieve as little as 0.04 ounces (some say as little as 0.008) of gold from each ton (2,000 pounds) of rock. Many contemporary gold-mining companies use this methodology, including the parent company of the 1980 corporation, Bodie Consolidated Mining. Bodie Consolidated formed to undertake a feasibility study for a cyanide heap-leach operation in Bodie in the late 1990s, but the operation did not come to pass. Cyanide heap-leach operations are often used in conjunction with open-pit mines.

Here's how the process works. First you build a *heap-leach pad* for processing the ore. The pad must be in a place that ore trucks can reach easily. It should be constructed at a slight angle to allow fluid running through the ore pile to drain in one general direction. The pad should be lined with a protective liner to guard against seepage and spillage.

Next you dig a pit to serve as a *cyanide pond,* to catch the metallic brew from the pad. You dig the pit where it will catch the pad's runoff, then line it with a protective liner, just as you did with the heap-leach pad. This pond collects an industrial-strength liquid that contains cyanide and precious metals, so it must be able to withstand brutal weather conditions and strong chemical wear and tear.

Now you heap crushed ore onto the pad and spray the heap with a cyanide solution continuously for months. The cyanide solution trickles down through the crushed ore, dissolving ("leaching out") the gold and other metals and taking them with it. When the metal-rich solution hits the pond, you collect the soup, separate the different metals out in your processing plant, and—voilà!—you have collected the gold. Not an enormous amount of it, and from a huge pile of rock, but after several huge piles of rock it starts to add up. You can reuse the cyanide solution, too, which helps keep costs down.

The economic efficiency of this process requires a large quantity of ore. The ore itself usually comes from the surrounding territory. Often the most efficient way to procure this is to blast a hole, collect up the rock debris, and process it.

This industry is not without its risks. Explosives must be handled carefully. Cyanide heap-leach operations use large quantities of toxic substances in what is usually a wilderness or semiwilderness environment and, correspondingly, require meticulous care and handling. When things go wrong, the price can be high for local residents of all species.

INDUSTRIAL ACCIDENTS

An example of an industrial accident involving a cyanide heap-leach operation can be found in Summitville, Colorado. The Summitville Consolidated Mining Corporation's heap-leach project went awry when its cyanide ponds leaked and/or overflowed, creating what U.S. Interior Secretary Bruce Babbitt termed an "extraordinary disaster" (Young and Noyes 1994). The U.S. Department of Health and Human Services' Agency for Toxic Substances and Disease Registry (ATSDR) Web site report and older U.S. Geological Survey reports explain what happened.[2]

When the open pit was dug, it exposed a lot of rock, whose minerals mixed with groundwaters that were rich in oxygen, making the resulting flow in nearby streams very acidic.[3] When the cyanide solution from the mining ponds spilled into the very acidic streams, some of the cyanide and acid reacted together to form hydrogen cyanide, which volatized into the

2. The ATSDR became involved under requirements set by the Superfund law, which requires its health assessment of proposed National Priorities list sites.

3. Please note that this occurs naturally in some mineral-rich areas when the rock and waters mix, without any mining activities. In areas with mining activities, however, operations sometimes increase the amount of rock mixing with water, thus increasing the effect in the area.

atmosphere. (Hydrogen cyanide is best known for its use in the execution gas chamber.) In summary, the concentrated chemicals (acid and cyanide), each of which causes its own damage, met and damaged each other, resulting both in more lethal and in less-harmful substances.

This was true when there was sunlight. When sunlight did not reach the water surface, such as during the snowmelt, more of the cyanide and metals made it downstream unchanged to negatively affect life and growth there. One of the most dramatic observations is that although the Summitville area was known through the late 1970s for its fishing, there hasn't been a fish in the area since 1990. The spill "has affected surface water quality for more than 20 miles downstream of the site" (U.S. Department of Health and Human Services 1997, p. 5). Summitville is currently undergoing cleanup operations.

The Summitville accident greatly alarmed those opposed to the Galactic mining development at Bodie. The proposed mining operations in Bodie were to be of the same type, and Galactic owned both the Summitville Consolidated Mining Corporation and the Bodie Consolidated Mining Company. Galactic, however, did not want to duplicate the problems of Summitville, which certainly didn't help profits. Galactic's resulting bankruptcy and sale of the Bodie property rendered the controversy irrelevant.

Our gold-mining and extraction techniques have changed since Agricola's day, yet some of the techniques from his time remain in use. As our technology increases the scale of our operations, it also increases the scale of potential damage in the case of accident. Today, an industrial mining accident can claim a multitude of lives—human and other—and permanently alter an established ecosystem. Mining remains a risky business, as it always has been, despite our technological advances.

T'a nee tsee na ah na ma ma no hopen,
t'a nee tsee ta nay tna a na ma ma no hopen . . .
—*"Mono Paiute Snowflake Song" welcoming*
the first high Sierra snow (Arlen, Batt,
Benson, and Kester 1995)

Appendix 3

The First Bodieites

THE KUZEDIKA

The first Bodieites arrived long before the gold-fevered prospectors. They were unconcerned about finding gold; other attributes of the area were far more important to them for their comfortable survival. Archaeologists estimate that humans arrived in the Bodie area about 5,500 years ago, and successive groups of people have kept the area inhabited, though sparsely, from then until now. Almost nothing is known about the very first inhabitants of the area. The Kuzedika had been there many generations by the time the first European-Americans arrived. Although they did not settle in the Bodie area specifically, they were in and around what is now the Mono County area.

GETTING BY IN THE EASTERN SIERRA

The Kuzedika crafted a surprisingly full lifestyle from the sparse and forbidding environment of the eastern Sierra. They worked hard at diverse strategies for survival, from hunting and gathering to a certain amount of agriculture.

Although the Mono County area was not an easy place to live, there was plenty for residents to eat when they knew where to look and at what time of year. Their diet had a wide variety; an overall list, in no particular order, would include *yuba* (a wild potato that tastes sweet when cooked), pine nuts, wild onions, *taboose* (which has a sweet, milky tuber or nut that grows at the base of its roots), Huki seeds, sweetgrass seeds, tule roots, California bulrush roots, swamp grass, wild mountain rice, elderberries, wild currants, buckberries, squaw cabbage, ryegrass seeds, sunflower seeds, cactus, trout, deer, antelope, rabbits, chipmunks, squirrels, gophers, porcupines, groundhogs, mice, quail, grouse, sage hen, ducks, geese, other edible birds, piaghi (Jeffrey pine caterpillars of the *Coloradia pandora* moth), and kutsavi (the hard-shelled larvae of the brine or alkali fly *Ephydra hians*). The Kuzedika also used salt scraped from alkali marshes.

They traded with neighboring people, some from the western side of the Sierra, for things they did not have. Many of their ancient traders' trails are trails to this day. They traded their obsidian, salt, rabbit-skin blankets, buckskins, pine nuts, kutsavi, piaghi, and pottery for such items as acorns, shell beads, bear skins, and black and yellow paints.

The Kuzedika moved about the region, harvesting crops as they ripened at different times in a given year. Generally, they ate some of the crop during the harvest, processed some for storage, and traded some for other goods. Likewise, they collected or hunted different animals, including edible insects and/or larvae, at different times of the year.

SPRING ONION HARVEST

In the spring, the Kuzedika harvested wild onions. They worked together in large groups digging a large pit and building a fire in it with wood from the "bull pine tree" (Jeffrey pine). According to Susie Jim, a Kuzedika interviewed in 1935, the pit would be "about eight feet around at the top, seven or eight feet deep and about six feet at the bottom." The fire burned all night long. In the morning, the people lay wet pine needles over the coals, then alternately layered onions and wet pine needles over and over again, finishing with pine boughs and dirt. After two nights, they took out the still-warm vegetables and pounded them into a dough, which they lay on pine needles to harden. According to Susie Jim, "They left them there for four or five days to dry out, until they became hard." They stored the hard vegetable dough in holes lined with grass and pine needles. They made soup from this hard dough by pounding it into flour, then putting the flour into a watertight basket and adding water, stirring until it made a thin soup. They often ate this with wild sunflower seeds.

THE KUZEDIKA AND THE KUTSAVI

Each year, the Kuzedika collected the kutsavi on Mono Lake. Every other year they also collected the piaghi, the caterpillars that appear biennially on the Jeffrey pine west of Mono Lake. They dried and stored both the kutsavi and the piaghi for later use and trade.

Old-timers said that each year the kutsavi began to appear in number at

Mono Lake in early July. Silas Smith, who was interviewed in 1935, when he was 76, claimed it was not only the Kuzedika who harvested the kutsavi, but also Indians from Benton, Bishop, Round Valley, and elsewhere.

As the wind blew clumps of kutsavi toward the shore, the women swept them into baskets using sweep brushes made out of rabbitbrush. If the kutsavi were not forthcoming, children would help the harvest along by kicking up the lake bottom, freeing the kutsavi from their underwater perches and setting them to float on the lake surface.[1] When their baskets were full, the women came ashore and poured the kutsavi out on a smoothed space to dry. Once the kutsavi were dry, they crushed them by hand, letting the wind blow the husks away. Some ate them just so; some cooked them into soup. "It tasted something like oysters," recalled Silas Smith.

The people stored the dried kutsavi in sacks made of "bitter brush bark."[2] They dug holes near cedar trees, put the sacks into the holes, and covered them with cedar bark and leaves, finishing up with dirt. The kutsavi that were stored this way were kept safe from spoilage and pilfering by animals.

Silas Smith said, "The young people now gather the kutsavi as the old people used to gather them and they still eat them," but he also commented, "[T]he kutsavi does not come out on shore now as thick as they used to." The harvest practice pretty much died out in the 1940s and 1950s, because of the dramatic reduction in the kutsavi supply. When Mono Lake's waters were diverted to the Los Angeles aqueduct, the lake became more saline (salty), which reduced the number of kutsavi.

THE PIAGHI HARVEST

Every other year, when the piaghi appeared on the Jeffrey pine trees west of Mono Lake, the Kuzedika harvested them as well. Susie Jim said the harvest time was right around the beginning of July. She described the harvest:

[T]he caterpillars . . . would start to come down off the trees. But before they start[ed] to come down, the Indians [would] dig trenches around the Bull pine trees. They were about two feet deep, around ten trees or more. These would belong to each camp owner. No one else could pick these caterpillars out of these holes but the owner of them. When the caterpillars started to come down they were green and white striped. They couldn't get out of these holes that the Indians had dug for them. They would start to come down in the morning when they got warm . . . they would keep coming down until about noon, then stop. Then the Indian women would start

1. Mono Lake Park Aide talk, September 1996.
2. This is as specific as Silas Smith was as to the type of bark used.

to pick them out of the trench into a basket. They went from tree to tree picking them up. . . . [T]hey smoothed off a place on the ground, took gravel and made it hot and took the caterpillars and poured them into this space. They took the hot gravel and parched the caterpillars with it. They gathered them up and separated the gravel from the caterpillars.

THE PINE NUT HARVEST

Fall was the time for the pine nut harvest. This was an important event because an abundant harvest helped to ensure there would be enough food stored up for the winter, which was generally severe. According to Silas Smith, during the harvest, the Kuzedika spread out among the piñon trees, working a large area. The men knocked the cones off the trees with a long stick, and the women collected them in baskets. When their baskets were full, they started a large pile either on the ground or in a larger basket. Not all these cones were open sufficiently to get the pine nuts out, but the people would either heat the cones in fire, which caused them to open up, or store them, still closed, for later use.

While some people gathered the cones, others prepared a portion of pine nuts for immediate munching by building a fire and tossing the cones into the fire to roast the nuts. They then broke open the cones, hulled the pine nuts, and ate them.

The Kuzedika method of storage protected the nuts from both animal marauders and spoilage. They left the nuts intact in the cones and packed them into a pile. They then put piñon branches over the pile and rocks over the branches, packing the rocks in well enough to keep the branches over the cones. They could take nuts as needed, repacking the branches and rocks each time.

The Kuzedika had two methods for roasting pine nuts: They either put the pine nuts in a hole and covered them with hot dirt (heated by fire) until they were done or put the nuts with hot coals into a particular type of basket and shook the basket, tossing the nuts up and down until they were roasted.

They hulled roasted nuts by cracking them with a rock, then tossing the nuts into the air, called "winnowing," using a large, flat basket. The wind blew the hulls away, leaving the nutmeats. The people then added more coals to the nuts, stirred them, and cleaned them after they were further cooked. Sometimes they stored roasted pine nuts in holes lined with cedar bark, which prevented the nuts from rotting.

The Kuzedika people made pine nut soup in a manner similar to that for their onion soup. First they pounded the nuts into flour on a flat rock, separating the heart from the meat, and then mixed the flour with hot water to make a soup.

Silas Smith asserted, "In the early days there used to be lots of pine nuts on the trees, but that was before the white man came to our country. After they came, the white man cut down the pine nut trees. Ever since then the pine nut trees do not bear so many pine nuts as they used to."

KUZEDIKA HUNTERS

The Kuzedika people hunted game large and small, using different techniques for different prey. They hunted both alone and in groups, again, depending on which animal was the quarry. They gathered together in groups for rabbit drives, burning the bush to flush the rabbits into waiting nets. Individual hunters trapped mice by using balanced stones with bait as a lure: The mice would seize the bait, causing a stone to fall upon them. Hunters followed porcupines to their dens and built a fire at the hole so the smoke would go inside, "where the porcupine was sitting. This way they smoked him to death," claimed Susie Jim. Then the hunter dragged the porcupine out, cleaned it, and put it into a fire to burn off the quills. They scraped off any remaining quills and cooked it. "How good it tasted!" Susie Jim remembered, "This was the Indians' food before the white men came." The Kuzedika people cooked most meat by baking it in a hole using hot dirt, although they roasted some over a fire as well.

The Kuzedika hunters also used bows and arrows. They made their bows of cedar, whittling them down and filing them smooth using rock files and scrapers, then coating them with a resinous glue that strengthened them. The bowstrings were made out of deer sinew. They made their arrows out of wild rose, "wild bamboo that grew around the springs,"[3] or willow. After cutting and filing the arrow, they made a hole in one end, put the flint in the hole, and poured hot pitch on the flint, holding it in place. They then tied deer sinew around the base of the arrow to further secure it. At the other end of the arrow they fastened three bird feathers, using deer sinew.[4]

KUZEDIKA CRAFT

The Kuzedika people used just about all parts of the animals they killed, not just the meat. The sinew went to bow and arrow making as well as for glue.

---

3. This is according to Silas Smith. Most likely the plant was *Purshia tridentata*.

4. The Paiute Shoshone Cultural Center in Bishop, California, features several examples of similar arrows.

They used the bones (and horns, if the animal had them) for tools and decorations and the brains, with water, to tan the hides. They made moccasins and suits out of tanned deer hides, and warm blankets from deer and rabbit skins.

They simply rolled out the deer-hide blanket to use it. Rabbit blankets were more complicated in their construction. Men and women working together cut the rabbit skins into strips, rolled them tight, and wove them together into a very warm blanket.[5] "It took about 20 rabbit skins to make a blanket," recalled Susie Jim. "[I]t took a whole day to make a rabbit skin blanket."

They made blankets out of willow bark by spreading the bark into layers large enough for a two-person blanket and tying them together: "The string they used to tie back with was the white bark off the same brush. . . . [W]hen finished, it was heavy, nice and warm," said Susie Jim. Willow bark was effective insulation. They used it to make shoes and "socks" as well as blankets. They wove the bark closely like a basket to fit their feet the way leather shoes do today. They made "socks" out of the same bark: not woven together, but rubbed down and laid in layers on the inside bottom of the shoe. The shoes laced up, and those who remembered them, such as Susie Jim, claimed, "These used to be lots of fun to wear when there was lots of snow in those days. These shoes were warm."

The Kuzedika made pottery and baskets, taking clay for the pottery from places such as Fish Springs, Baker Creek, and Coso Hot Springs, to the south. Their baskets, which to this day are known for their fine quality, were generally made of willow.[6] Kuzedika women made different baskets for different purposes. There were watertight baskets for carrying water and for cooking, baskets for tossing seeds to rid them of their hulls, baskets on boards for carrying babies, and so on.

ROUND HOUSES

The Kuzedika built what they called in English "round houses." They dug a round trench, and in it placed poles made from trees, leaning them together at the top to form a wide cone. They added cedar bark until they had covered the outside and there were no cracks left. The women then poured pine needles around the outside of the house to cover it thickly, beating the needles down with a stick until it was solid. They made a hole in the top for smoke to escape, built a fire in the center of the house, and called it home. Despite the

5. The Bishop Paiute Shoshone Cultural Center features an excellent display of this technique and materials.

6. There are several impressive examples of these baskets in the Bridgeport museum.

lack of modern insulation and building codes, the houses were very warm and snug, even in the snow.[7] (A Kuzedika round house in the snow appears in chapter 5.)

THE KUZEDIKA SOCIETY

The Kuzedika developed their lifestyle and traditions over years in a climate of dramatic seasonal changes. Their customs provided for constant renewal of their crucial resources. Each generation maintained the tribal social structure. Men and women had their well-defined, different duties within the group. Family units were respected within tribes. Children were instructed by their families through stories handed down from their elders. There were celebrations and rituals surrounding many aspects of life. These rituals, still known by some contemporary Kuzedika, are held dearly and respectfully, and are kept private.

The Kuzedika people had their own doctors. As Susie Jim explained,

The old Indians believed in the Indian Doctor or the medicine man, they do so yet. When the Indians got sick they would send for an Indian Doctor, whether it was a man or woman that was sick, to come and doctor them. The doctor always came in the evening to do his work, from night time until daylight. They sang and danced around the big fire before starting to do his work. He sat down by the sick person and asked how they became sick. . . . [T]he doctoring always took place outside in the open air.

According to Susie Jim, the Kuzedika approach to medicine revolved around dreams and purification. The patient would relate the dream that brought the sickness. Then the doctor would sing his medicine song, go around the fire, and, placing his mouth where the patient's pain was, draw the sickness out. He repeated this several times until he had figured out the origin of the trouble, which he would then relate to the patient.

THE KUZEDIKA AND OTHER PEOPLE

[S]ince we are surrounded by inaccessible rocks and precipices, we have hitherto been exempt from the rapacity of the nations of Europe who have an inconceivable lust for the pebbles and mud of our land and would kill us to the last man to get possession of them. —Old man in Eldorado, Voltaire (Candide, circa 1759)

---

7. Examples of a Kuzedika house can be found in the Mono Lake Visitor Center and the Bishop Paiute Shoshone Cultural Center (in Bishop, California). See a photograph of it in chapter 5.

The harshness of the local environment meant that few others competed with the Kuzedika for possession of such an area. Prior to 1800, some Indian tribes in parts of what is now the western United States had had little or no contact with European-Americans:

These were the ill-defined groups, such as the Paiute, who lived in the quite inhospitable Great Basin region. That harsh and stingy environment had long forced them to live the simplest of Indian lifestyles as they moved constantly in search of food. Additionally, their lot had worsened by 1800 because Plains, Plateau and Southwest marauder Indians all raided the Great Basin looking for captives to trade for Spanish and American goods. Nevertheless, in 1800 nobody desired their land, and they would not be elbowed aside by covetous Americans for half a century. (Champagne 1994, p. 206)

In the mid-nineteenth century, when the first pale intruders descended on the Kuzedika and nearby Paiute tribes, many chiefs of the time at first welcomed them respectfully. Some, such as Paiute Chief Truckee, Sarah Winnemucca's grandfather, were even joyful, viewing the arrival of the whites as the reunification of the family of humankind and fulfillment of an ancient legend.

The settlers in the Bodie area, however, had no such reunion on their minds; they were coming to find their fortunes in gold or to take advantage of the frontier with its wide, open, "unclaimed" spaces. There was minimal concern for the people who had been living there for centuries. The fact that the Kuzedika looked and lived differently from the settlers, didn't speak the same language, and had inferior weaponry (in particular, no guns) made it easier for the European-Americans to dismiss the notion that the Kuzedika had any claim to the land. "They came like . . . a roaring lion, and have continued so ever since, and I have never forgotten their first coming," recollected Sarah Winnemucca ([Sarah Winnemucca] Hopkins [1883] 1969).

The Kuzedikas' very survival depended on regular access to the different plants and animals in the local ecosystem, which covered a large geographic area. When the Euro-Americans entered the picture, especially the cattle ranchers and farmers, they threatened those resources and restricted access to them. Suddenly the land on which the Kuzedika had always hunted and gathered was full of cattle. Cattle were incompatible with the indigenous crops and animals in many cases. Cattlemen were unenthusiastic about people taking their cattle for food without paying for it, even if those people were starving because those same cows had destroyed their usual food supply. Conflict was almost guaranteed.

The Kuzedika people were soon overruled in land-use and -ownership matters, as were most if not all California Indians. "Wherever whites had demanded Indian land, sooner or later the Indians were forced to cede their holdings" (McGrath 1984). This might not have happened (or at least not so soon) if it were not for the superiority of the Americans' weapons. Stephen Powers, in his 1877 tome, *Tribes of California* (republished in 1976), commented,

> Let it be remembered . . . that after the Republic had matured its vast strength and developed its magnificent resources, it poured out hither a hundred thousand of the picked young men of the nation, unincumbered [*sic*] with women and children, armed with the deadliest steel weapons of modern invention . . . and pitted them against a race reared in an indolent climate, and in a land where there was scarcely even wood for weapons. . . . Never before in history has a people been swept away with such terrible swiftness, or appalled into utter and unwhispering silence forever and forever, as were the California Indians by those hundred thousand of the best blood of the nation. . . . Let a tribe complain that the miners muddied their salmon-streams, or steal a few pack-mules, and in twenty days there might not be a soul of them living. (p. 404)

Editor Robert Heizer notes in his introduction to *Tribes of California,* "By 1870, from fifty to seventy thousand Indians were blown away by the well-armed Americans and by starvation and disease. Even the will to live had been destroyed."

There were conflicts in the eastern Sierra, mostly south and north of Bodie, in the second half of the nineteenth century. Eventually, reservations were established to the south and north of Bodie. The Kuzedika people were in between areas of major conflict and as such were not specifically targeted for removal by the U.S. government. But their closest neighbors in the South were.[8] And many children in the region were sent away from their families to Indian schools as recently as the first half of the 1900s. When the town of Bodie formed and grew, there were several Kuzedika Bodieites, most of whom lived in their own section of town.

8. A detailed description of what it was like to be forced to walk south, down California's interior, at the behest of the U.S. government, can be found in the interviews of Bishop Paiute people in the Ethnological Collection at the Bancroft Library, University of California at Berkeley.

Oh, what was your name in the States?
Was it Thompson or Johnson or Bates?
Did you murder your wife and fly for your life?
Say what was your name in the States?

> —Gold Rush–era song
> (Arlen, Batt, Benson, and Kester 1995)

Appendix 4

When Bodie Meets a Bodie

WAITING FOR BODIE

There are facts, both supportive and contradictory, for each of the W. Bodies who have been nominated as Bodie's founding father. As of yet, no individual has been proven to be him. A perusal of the available records of the day yields the following information.

WAKEMAN BODEY

One book, published in the late 1900s, refers to Wakeman Bodey as the founding father of Bodie and asserts that he hailed from Poughkeepsie, New York (Loose [1979] 1989). Federal census records from 1850 indicate that a Wakeman L. Bodey lived in Poughkeepsie, New York,[1] and worked as a tinsmith. Wakeman L. Bodey was born in America in 1801 or 1803. His wife, Sarah, a seamstress, was born in America in 1811. They had two children, Ogden E., born in 1846, and Mary A., born in 1848 (U.S. Federal Census 1850). Wakeman Bodey appears in the 1860 census as still being in Poughkeepsie. Although census records from those days are notoriously unreliable, if this is accurate he cannot be Bodie's founder. Two other strikes against him are his middle initial, which is L., not S., and the fact that in 1859 he would have been fifty-eight or fifty-six years old, at least ten years older than his friends in the newspapers described him.

---

1. Poughkeepsie, New York, is located in Duchess County. Could this be the origin of the notion that Bodie was Dutch?

William Smith Bodie appears in the 1850 New York census, as does Wake-man L. Bodey. William Smith Bodie was a farmer in Hammond, New York. He was born in Scotland in 1812, which would make him forty-seven in 1859, much closer to the agreed-on age than Wakeman L. Bodey. Bodie's wife, Elizabeth, was born in America in 1820. They had two children, Mary, born in 1843, and William, born in 1849. Neither William Smith Bodie nor his family appears anywhere in the 1860 federal census or New York census.

The strikes against William Smith Bodie are that he was not from Pough-keepsie, which was the impression some people had (as stated in the Bodie papers), and the disappearance of his family from the census rolls of 1860. The arguments in favor are that his initials are correct and he came from Scotland, furthering the Scottish-origin theory. Apparently, the name Bodie (spelled thus) is associated with several families named Smith who come from the region of Buckie, Scotland.

Gordon Bodie Smith, another relative of the Scottish Bodie, says his family records suggest this relative of his might have been Bodie's founder (Smith 1996). William Smith Bodie's family name was always thus spelled. He left his native Buckie, Scotland, on a family ship, and at some point made his way to Monoville, they are certain, because the family received letters from him when he was there. The family lost touch with him after receiving these letters, however. Relatives of the Scottish Bodie speak of claim letters they received from the U.S. government when Bodie died, which could be further indication of his identity. Unfortunately, such letters are hard to find 150 years later. The search for further information is ongoing.

In the late 1920s, J. S. Cain received several inquiries from individuals who were convinced they were Bodie's rightful heirs. Several of these inquiries appeared to be from different members of the same family; some were still in Scotland, and some were in the United States. Cain received letters from individuals in Los Angeles, California; Salt Lake City, Utah; Golden, British Columbia; Oakland, California; Washington, D.C.; Buckie, Banffshire, Scotland; and Edinburgh, Scotland. A few of the correspondents claimed that Bodie had been murdered, something that does not appear in other sources. Several of the correspondents were interested in any inheritance they might have coming to them, but Cain was unable to help them. As he noted patiently in reply after reply, "Referring to William Smith Bodie, there is no record in Mono County of title to any property in Bodie."

Neither William Smith Bodie nor Wakeman L. Bodey appears in the 1840

New York census. The fact that they both appear in the 1850 census would tend to discredit the claim that our man Bodie rode the *Matthew Vassar* into San Francisco in 1848. There is no mention of either Bodey or Bodie in the California census for 1850 or 1860.

THE WILLIAM BODYS

The 1850 California census roll includes a William Body and a Willis Body, and the 1860 California census roll includes two William Bodys. In 1850, both William and Willis Body were mining in Mariposa County, but they were very young, twenty-one and eighteen, respectively. In 1860, both William Bodys were living in Sacramento County. One was a twenty-eight-year-old steward from New York, the other a twenty-two-year-old porter from Virginia. Both of these Bodys were younger than the town founder and were noted as being black men, which it is assumed Bodie was not, merely because he was never noted as being so by any who knew him, and it would not have escaped noting in those days.

Which Bodie is the real McCoy? The truth at this point belongs to the ages. But the lingering mystery only typifies Bodie's history, which is one of uncertainty and exaggeration.

❧ NOTES

1 · WELCOME TO BODIE

1. Kuzedika was a specific name of the Mono Paiute Indians, just as the Honey Lake Paiutes were the Wadatika, the "wada-seed eaters," and the Pyramid Lake people were the Kuyui-dika, or "sucker-eaters."

2. The Ahwahneechi are also known as the Yosemite Indians.

3. As a side note, the change in spelling from the original "Yosemity" to "Yosemite" is generally credited to Lieutenant Moore's 1851 published account of this expedition (Bunnell 1892; Russell 1928). The name *yosemity* or *yosemite* is an Ahwahneechi word meaning "full-grown grizzly bear."

2 · THE GOLD IN THEM THAR' HILLS

1. As noted earlier, the Mining District's name was originally spelled Bodey. It was not until November 1862 that the first references to the town and district were spelled *Bodie* (Mono County Records, Bodey Mining District, Book A, 1862).

2. The Isabella tunnel was known in later days as the Syndicate.

3. Aurora was originally thought to be in California. In fact, it was the original location of the Mono County seat. Imagine everyone's surprise when the 1863 boundary survey party determined that Aurora was actually in Nevada by "some three and one-third miles." It had been a topic of debate for some time, and the survey finally put the issue to rest.

4. In some records his name is spelled Eshington.

5. In 2002 prices, that would be almost $11.5 million.

6. In 2002 figures, it would be almost $49,110 per ton. At the end of the twentieth century, it was common for mining companies to use cyanide technology on ore valued at less than $10 per ton.

7. None of the Bodie papers of that time that are available today (see the references) featured the oft-quoted rewrite "Good, by God! We're going to Bodie!" This is not to say that the rewrite didn't appear, only that it was not in any of the papers still available today (Shipley 1995).

3 · THE LOUSY MINERS

1. Jones splitters are devices that split the flow of rock so that half goes out each side of the splitter.

2. Chipmunk crushers are also called "jaw crushers," because they crush rock with a movement suggestive of jaws. Often the term "chipmunk crusher" is used to refer to a smaller jaw crusher.

3. This mesh is a screen with 150 openings per square inch.

4. This means that he needed to test, or assay, 29.166 grams of this pulp to correspond to an entire ton of ore. That's why it's called an "assay ton."

5. Flux is a dissolving agent. It dissolves particles that would otherwise keep the reaction from happening.

6. It is called a muffler because the flame doesn't go directly onto the material being assayed, but is muffled around a compartment, which gets hot inside.

7. Generally at this point in the assay process a reducer of some sort is added to the mixture if the ore is nonreducing. Common choices of reducers include charcoal, flour, sugar, starch, argols, or sulfur. If the ore contains natural sulfide, the reducer is unnecessary. If the ore is *too* powerful as a reducer, an oxidizer may be added (*Collier's Encyclopedia* 1950, p. 383).

8. The percentage of gold in the bars varied widely, depending on which mine it came from.

9. The other Jupiter men were Joe McDonald, Barney McDonald, George Harber, Joseph Burnett, James Murphy, and William Andrews.

10. In late October of 1879, Daly represented the Bodie and Mammoth mining districts in the financial markets of New York, a job he performed quite well. The next spring he was appointed to a superintendent post in Leadville, Colorado. He remained a celebrity in Bodie, and the papers followed his exploits. Daly was eventually killed in a skirmish with Apache people in New Mexico in 1881.

4 • THE GOLDEN TIME

1. The Pacific Stock Exchange Inc. was descended from the San Francisco Stock and Exchange Board, which officially opened in September of 1862. The Pacific Exchange closed 140 years later, in 2002.

2. Presumably, Captain Gregg was the popular fellow responsible for the street watering, but no further information was available.

3. McGrath makes this staggering claim based on FBI-style statistical analysis of recorded murders. He found that only comparable frontier towns' murder rates (such as Virginia City) came anywhere near that of Bodie's. McGrath's analysis entailed dividing the number of murders by the span of years in which they took place (five years for his research), then multiplying the resulting number by twenty, "to convert Bodie's average population for that time to the 100,000 population norm used by the FBI Crime Index" (McGrath 1984).

5. BODIE'S SOCIAL LADDER

1. The Red Light Abatement Act declaring brothels public nuisances passed in 1913, after many years of national debate over how to deal with prostitution (Bean 1968). The argument over whether regulation or legislation is the best approach continues to this day.

2. Some sources note that there was concern about the prescription mistake (was it error? intentional? suicide?), but nothing appears to have come of it. Her death was ruled accidental (Seagraves 1994).

3. The final verdict was not found in the resources available.

4. Bodie's female suicide rate in the boom years was 100 per 100,000, compared to the 1978 U.S. rate for women of 6.3. Women in Bodie were six or seven times more likely to commit suicide than men. McGrath (1984) goes into detail on why this was, but there are recognized influential factors. The women in Bodie were quite isolated compared to women in other regions of the United States. There were very few women in Bodie (only 10 percent of the boom-time population) to begin with, and many of these were prostitutes. Prostitutes led very difficult lives, often with poor health, poor treatment from the surrounding society, and few things to look forward to. Many became addicted to opium as well. All these factors influenced the situation that made the suicide rate for women so much higher than that for men in Bodie. The opposite was true elsewhere then and now. In 1870s Philadelphia, for example, men were five times more likely to do themselves in than were women. In the United States as a whole, in the 1950s, 1960s, and 1970s, men were three times more likely. These figures were derived using the same statistical methodology as that used for the homicide rate (McGrath 1984).

5. The 1880 census notes residents from Ireland, Canada, England, Wales, China, Germany, Scotland, Mexico, France, Sweden, and Norway.

6. Both were culinary staples of the Kuzedika. Kutsavi are the larvae of the brine fly, and piaghi are caterpillars that favored the nearby Jeffrey pine trees. For more details, please see appendix 3.

7. This was a term commonly used in the mainstream press to denote Indians of any tribe.

8. James Fenimore Cooper, who wrote such works as *The Deerslayer* (1841).

9. In fact, the Chinese return rate between 1848 and 1882 was about 47 percent, compared to 30 percent for the British, more than 60 percent for the Italians, and 33 percent for the Japanese (Chan 1990, quoted by Yung 1996).

10. This practice continued in California until about 1949 (Yung 1996).

11. "Celestials" is a term commonly used in the mainstream press of the day to denote Chinese people; it is derived from references to China as the "celestial empire."

12. "John" is a term commonly used in the mainstream press of the day to denote Chinese people; it comes from the tendency for Europeans and European-Americans to refer to Chinese men as "John Chinaman" because the Anglos could not easily remember and pronounce proper Chinese names and because Anglos often perceived Chinese men as indistinguishable from each other.

13. The law's target was actually the white population; it prohibited white people from marrying nonwhite people. Such marriages were not illegal in Nevada, however, and several couples married in Nevada and then lived in California.

14. Prostitution itself was not outlawed until 1913 in California, so that some interpret this law as aimed more at Chinese women than at the profession of prostitution.

15. The Scanavinos owned Goat Ranch, off the Cottonwood Canyon road about six miles or so out of Bodie. Many deceased Scanavinos are buried in the Bodie cemetery.

16. Joss temples got their name from the "joss sticks" or incense sticks that were burned there.

1. Both Treloar and DeRoche first appear in the Mono County Register in 1879. They were noted respectively as Thomas Treloar, a thirty-one-year-old English miner, and Jules Daroche, a forty-eight-year-old French Canadian carpenter. Johanna Lonahan was also said to be newly arrived in 1879, although women were not recorded in the County Register (Mono County Records, *Great Register of Mono County* 1879).

2. The Treloar house was located on Mills Street below Lowe Street, one block west and one block south of the Miners' Union Hall, between what is today the SHP county barn and entrance kiosk.

3. Meaning "full of smoke or soot."

4. At the 2002 gold price of $310.90 per ounce obtained from http://finance.yahoo.com, that would be $216,663,599.06.

5. For a point of reference, in the 1990s the population of Bridgeport, the Mono County seat, was under 1,000, closer to 500.

7 • UP HOME

1. "High grading" is thievery of ore by miners before it reaches the mill.

2. Dr. Robert W. Sprague recalls a coworker's tales from the 1950s: "One of the guys I worked with . . . was trying to get a group together to fund a diving expedition to Bodie to retrieve sacks of high-grade ore he had hidden in a side tunnel in one of the mines before it shut down. Of course it was under water . . . so they would need diving equipment. . . . I don't think anything ever came of it."

3. Sadly, the plates have been stolen now, presumed to have been pilfered for their brass.

4. In his book *Mining Camp Days* ([1968] 1986), Emil Billeb described in interesting detail a visit in San Francisco from Sam Leon and his Kuzedika wife and children.

5. Homestake leased its claims to Galactic Resources Limited (or its subsidiary, Bodie Consolidated Mining Company) in the late 1980s. About the same time, Homestake became a majority shareholder of Galactic, making an interesting Celtic knot of mining interests and claim ownership in the Bodie area.

6. For the uninitiated, such as I was when I started this book, a "thundermug" is a chamber pot.

7. In 1945, after the Roseklip operations had ceased, another fire burned more of the town and the cyanide plant. The Bodieites tried to use the water from the storage pond to fight the fire, but the trout in the pond clogged the outlet and the water wouldn't flow.

8. What Peterson is referring to relates to the War Powers Act of 9 March 1933, with which President Roosevelt halted mining activities. It stated, among other things, "the President may . . . prohibit under such rules and regulations as he may prescribe by means of licenses or otherwise, any . . . export, boarding, melting or earmarking of gold or silver coin or bullion or currency by any person within the United States or anyplace subject to the same jurisdiction thereof."

9. This means the last time you could get a Bodie postmark was in November 1942.

1. It is presumed to be Ella Cain who had the springs under the floor removed in the hall. What was a distinct advantage for dances was a liability in a museum.

2. Yes, that does imply a total of between 2.3 and 4.6 million ounces of gold, in contrast to the stated expectation of 1 to 1.25 million ounces. This contradiction remains unresolved.

3. Further research shows that what went around came around, because according to the 14 August 1994 *New York Times* (Young and Noyes 1994), Homestake Mining became Galactic's largest shareholder in 1988.

4. The Nevada Division of Wildlife's Web site specified that these mortalities "may result from cyanide, or other types of chemical poisoning," although cyanide is agreed to be the major concern where cyanide heap-leach operations are under way. The good news is that although the year of 1986 saw 2,000 animal deaths alone, 1993 and 1997 had fewer than 400. This is likely due to improved methods of repelling animals from the ponds. (See Nevada Department of Wildlife [NDOW] Mining Program, at http://dcnr.nv.gov/nrp01/reso5.html.)

5. When SCMCI, Galactic's Summitville subsidiary, gave the state of Colorado its required report of projected cleanup costs for its Summitville site, the cost projected amounted to "at least $20 million. . . . Three days later Galactic announced it was filing for bankruptcy" (Obmascik 1993a).

REFERENCES

Adair, Katie Conway Bell. 1987 (October). Interview by Bodie SHP volunteer Diana Mapstead. Oral History Collection, Bodie SHP, Sierra District, California Department of Parks and Recreation. Tape recording.

Agreement spares Bodie from threat of development. 1998 (22 December). *Reno Gazette-Journal.*

Agricola, Georgius. 1556. *De re metallica.* Basel: J. Frogen and N. Episopius.

Ancient Order of United Workmen of Bodie. 1876-1912. AOUW Lodge No. 143 records (BANC MSS C-G 107). Bancroft Library, University of California at Berkeley. Manuscript.

Arlen, Karen, Batt, Margaret, Benson, Margaret Ann, and Kester, Nancie. 1995. *They came singing: Songs from California's history.* Oakland, CA: Calicanto Associates.

Aurora Esmeralda Union. 1867 (23 November-28 December).

——. 1868 (4 January-22 February).

——. 1878 (May-July).

——. 1880 (19 April).

Ball, Howard T. 1977 (October). Interview by Ranger Ken Featherstone. Bodie SHP. Oral History Collection, Bodie SHP, Sierra District, California Department of Parks and Recreation. Tape recording.

Bancroft, Herbert Howe. 1888. *The works of Herbert Howe Bancroft: History of California,* Vol. 6. San Francisco: History Company Publishers.

Bean, Walton E. 1968. *California: An interpretive history.* San Francisco: McGraw-Hill.

Beasley, Delilah. 1919. *The Negro trailblazers of California.* Los Angeles: Times Mirror Print and Binding House. Bancroft Library, University of California at Berkeley.

Bell, Bob. 1986 (August). Interview by Ranger Jack Shipley. Bodie, CA. Oral History Collection, Bodie SHP, Sierra District, California Department of Parks and Recreation. Tape recording.

——. 1994 (December), 1996 (August). Conversations with the author. Luning, NV, and Bodie, CA, respectively.

Bell, Gordon, and Bell, Jeanne. 1994 (December). Interview by the author. Bridgeport, CA.

Benton Mono Weekly Messenger. 1879 (1 February-22 March).

——. 1880 (5-19 April).

Benton Weekly Bentonian. 1879 (16, 21 August).

——. 1880 (22 January; 14, 28 February; 19-26 April; 29 September).

Billeb, Emil. [1968] 1986. *Mining camp days.* Las Vegas: Nevada Publications.

Blasts blamed for damage to historic house. 1991 (18 January). *Reno Gazette-Journal.*

Bodie. 1952 (December). *Westways,* 18: 4-5.

Bodie Chronicle. See *Bridgeport Chronicle-Union.*

Bodie Consolidated Mining Company. 1989 (22 August). Correspondence with Richard Rayburn of the California Department of Parks and Recreation.

Bodie Daily Standard (a.k.a. *Bodie Standard, Standard, Bodie Weekly Standard, Daily Bodie Standard,* and *Weekly Standard-News*). 1877 (November–December).

——. 1878 (February–March; 6-11 July; 10-17 September; December).

——. 1879 (January, February, March, May, June, July, December).

——. 1880 (January–December).

——. 1881 (15 July).

Bodie district adds a new chapter to its history (The). 1989 (February). *Engineering and Mining Journal,* 190(2): 7.

Bodie Evening Miner. 1890 (4 August).

Bodie Miner. 1890 (4 August).

——. 1892 (22 April).

Bodie Miner Index. 1897 (23 January; 13 November).

——. 1898 (15 October).

Bodie Morning News (a.k.a. *Daily News*). 1879 (2-31 July; 2-29 August; 2, 6-10, 12-13, 17, 19-28 September; 2-5, 8-31 October; 2-26, 29 November; 12, 19-31 December).

——. 1880 (1 January–18 March; 11, 13 May; 4, 8-11, 14-16, 19, 22, 24-28 June; 22, 24 July.

Bodie park land deal OK'd in BLM buyout. 1999 (7 January). *Mono County Review Herald.*

Bodie park will double its acreage. 1997 (19 June). *Mono County Review Herald.*

Bodie Standard. Also known as *Bodie Daily Standard, Bodie Weekly Standard, Daily Bodie Standard, Standard,* and *Weekly Standard-News.*

Bodie Weekly Standard. Also known as *Bodie Daily Standard, Bodie Standard, Daily Bodie Standard, Standard,* and *Weekly Standard-News.*

Bridgeport Chronicle-Union (a.k.a. *Bodie Chronicle, Bridgeport Union, Mono-Alpine Chronicle*). 1878 (December).

——. 1880 (May–June; 7, 10, 24 July; 21, 28 August).

——. 1881 (15 January).

——. 1884 (1-18 March).

Bridgeport Union. See *Bridgeport Chronicle-Union.*

Browne, John Ross. [1865] 1981. *A trip to Bodie Bluff and the Dead Sea of the West (Mono Lake) in 1863.* Golden, CO: Outbooks.

——. 1869. *Adventures in Apache country.* New York: Harper and Brothers.

Bryant, Marian Hitchens. n.d. Letter to Bodie SHP. Unit History Collection, Bodie SHP, Sierra District, California Department of Parks and Recreation.

Bryson, Jack. 1973 (July). Interview by Ranger Mark Pupich. Unit History Collection, Bodie SHP, Sierra District, California Department of Parks and Recreation.

Bunnell, Lafayette H. 1892. *Discovery of the Yosemite and the Indian war of 1851.* New York: Fleming H. Revell Company.

Cain, Ella. 1956. *The story of Bodie*. San Francisco: Fearon Publishers.

———. 1961. *The story of early Mono County*. San Francisco: Fearon Publishers.

Cain, James Stewart. 1927. Typed correspondence with Gretta Brooks. Sacramento: California State Library.

———. 1930. Typed correspondence with James Nicholson. Sacramento: California State Library.

Cain, Sadie, and Cain, Stuart. 1977 (3 August). Interview by Bodie SHP staff member. Unit History Collection, Bodie SHP, Sierra District, California Department of Parks and Recreation.

Calhoun, Margaret. 1984. *Pioneers of Mono Basin*. Lee Vining, CA: Artemisia Press.

Carpenter, Harold Brady. 1881 (23 January). Letter to Frank Marion Gilcrest. Unit History Collection, Bodie SHP, Sierra District, California Department of Parks and Recreation.

Chalfant, William. 1928. *Outposts of civilization*. Boston: Christopher Publishing House.

Champagne, Duane. 1994. *The Native North American almanac*. Detroit, MI: Gale Research.

Champion, Dale. 1989 (26 June). Gold hunt jostles ghost town. *San Francisco Chronicle*.

Chan, Sucheng. 1990. European and Asian immigration into the United States in comparative perspective, 1820s to 1920s. In Virginia Yans-McLaughlin, ed., *Immigration reconsidered: History, sociology and politics*. New York: Oxford University Press.

Chappell, Maxine. 1947 (September). Early history of Mono County. *California Historical Society Quarterly*, 26(3): 233-248.

Chesterman, Charles, Chapman, Rodger, and Gray, Cliffton, Jr. 1986. *Geology and ore deposits of the Bodie Mining District, Mono County, CA*. Bulletin 206. Sacramento: California Department of Conservation, Division of Mines and Geology.

Clough, E. H. 1878 (1 June). The bad man of Bodie: A mining expert meets Washoe Pete. *San Francisco Argonaut*, 2: 7.

Colcord, R. K. 1928 (June). Reminiscences of life in territorial Nevada. *California Historical Society Quarterly*, 7: 112-120.

Collier's Encyclopedia. 1950. 2:383. New York: Collier and Son.

Daily Bodie Standard. 1878 (6-11 July; 21 September; 10-17, 21-30 December).

———. 1879 (20, 21, 29-31 January; 4-8, 11-17, 19-20, 24, 26, 27 February; 14 March; 24 May; 12, 13, 24, 27 June; 10, 12-19, 24 July; 1-13, 15-19, 21, 25 August; 2, 5-12, 16-30 September; 3, 6-30 October; 1-20, 22 November; 9, 11, 13-31 December).

Daily Free Press (a.k.a. *Free Press*). 1879 (3, 4, 6-15, 19-24, 26 November; 6, 13, 16-26, 29, 30 December).

———. 1880 (2, 5-22, 24 January; 10, 12 February; 5, 7-25, 28-31 March; 2-9, 11, 14-16, 18-24, 28 April; 27, 29 May; 1-10, 12-17, 19-25, 30 June; 14, 16, 17, 20-22, 25 July; 25, 27-30 November; 2-31 December).

———. 1881 (1 January; 4, 6-9, 12, 18, 20-22, 24 February; 5, 8-16 March; 1-9, 12 April; 25, 27, 28 May; 3, 5-30 June; July-December).

———. 1882 (3-10, 12 January; 31 May; 1 June; 7, 10-18, 20-31 October; November, December).

———. 1883 (January-April).

Daily News. See Bodie Morning News.

Davis, Dorothy Marie. 1932 (December). Ghosts still walk in Bodie. *Overland Monthly,* 90: 299-300, 312, 316.

deDecker, Mary. 1966. *Mines of the eastern Sierra.* Glendale, CA: La Siesta Press.

DeLyser, Dydia. 1995. Ghost town story. *Big important town.* Syracuse, NY: Syracuse University Press.

Dillinger, Bill. 1989 (26 March). Is this ghost town giving up the ghost? *Sacramento Bee.*

Dodsley, Robert. 1817. *The economy of human life.* Philadelphia: Edward Earle.

Fletcher, Thomas C. 1987. *Paiute, prospector, pioneer.* Lee Vining, CA: Artemisia Press.

Frickstad, Walter N. 1955. *A century of California post offices, 1848-1954.* Oakland, CA: Philatelic Research Society.

Gandosi, Bruce. 1996. Telephone conversation with the author.

Ghost towns of California: Bad men from Bodie. 1952 (7 January). *Fortnight,* 12: 27-28.

Giffen, Guy J. 1940. Six shooter's who's who (BANC Mf591.G5). Bancroft Library, University of California at Berkeley. Manuscript. Photocopy.

Goodwin, Ed, and Dolan, Pat. 1994 (December). Interview by the author. Oakland, CA.

Gray, Lauretta Miller. 1994 (December). Interview by the author. Oakland, CA.

Gray, Lauretta Miller, and Tracy, Fern Gray. 1988. Interview by Frank Lortie, Historian. II. Unit History Collection, Bodie SHP, Sierra District, California Department of Parks and Recreation. Typescript.

Hanson, Earl. 1981. *Understanding evolution.* Oxford: Oxford University Press.

Hart, John. 1996. *Storm over Mono.* Berkeley: University of California Press.

Hillinger, Charles. 1989 (4 June). Mining firms strikes a deep vein of concern for historic ghost town. *Los Angeles Times.*

Hogle, Gene. 1943 (July). Bad man from Bodie. *National Motorist,* 20: 17-18.

Hopkins, Sarah Winnemucca. [1883] 1969. *Life among the Piutes: Their wrongs and claims.* Bishop, CA: Chalfant Press.

Huston, Ann, and Tilghman, B. Noah. 1997. Preserving a historic mining landscape. *Cultural Resource Management,* 20(9): 41-45.

Independent Order of Odd Fellows of the State of California. Bodie Lodge Odd Fellows records. 1876-1916 (BANC MSS C-G 106). Bancroft Library, University of California at Berkeley. Manuscript.

International Union of Mine, Mill and Smelter Workers, No. 61, Bodie, CA. Bodie Miners' Union records. 1890-1913 (BANC MSS C-G 109). Bancroft Library, University of California at Berkeley. Manuscript.

———. Fraternal Burial Association Records. 1898-1908 (BANC MSS C-G 108). Bancroft Library, University of California at Berkeley. Manuscript.

Jim, Susie. 1935. Interviews with unnamed anthropologists. Oral History Collection, Bancroft Library, University of California. Microfilm.

Johnson, Russ, and Johnson, Anne. 1967. *The ghost town of Bodie, a California state park*. Bishop, CA: Chalfant Press.

Johnson, Ted. 1991 (22 May). Mines, homes find no peace as neighbors. *Los Angeles Times*.

Joseph, Dorothy. 1987 (October). Interview by Bodie SHP volunteer Diana Mapstead. Oral History Collection, Bodie SHP, Sierra District, California Department of Parks and Recreation. Tape recording.

Keely, C. C. 1925. Unit History Collection, Bodie SHP, Sierra District, California Department of Parks and Recreation. Typescript.

Knickerbocker, Brad. 1991 (20 March). Gold rush poses threat to land. *Christian Science Monitor*.

Kobashigawa, Dr. Ben. 1996 (15 May). Telephone conversation with the author.

Kriel, Zady. n.d. Letter to Bodie SHP. Unit History Collection, Bodie SHP, Sierra District, California Department of Parks and Recreation.

Lafee, Helen. 1973 (25 June). Letter to Bodie SHP. Unit History Collection, Bodie SHP, Sierra District, California Department of Parks and Recreation.

Lavender, David. 1944 (August). An old ghost stirs. *Pony Express Magazine*, 11: 3–6.

Lawrence, Steve. 1989 (18 June). Bodie's gold: New interest may threaten California ghost town. *Vallejo Times-Herald*.

Lee, H. D. 1996. Telephone conversation with the author.

Leeds, Esther. 1986 (14 January). Letter to Bodie SHP. Unit History Collection, Bodie SHP, Sierra District, California Department of Parks and Recreation.

Levy, Jo Ann. 1992. *They saw the elephant: Women in the California gold rush*. Norman: University of Oklahoma Press.

Lewis, Arena Bell. 1995 (March). Interview by the author. Yerington, NV.

Loose, Warren. [1979] 1989. *Bodie bonanza*. Las Vegas: Nevada Publications.

Lundy Homer Mining Index. 1881 (15 January).

MacFarren, H. W. 1914. Practical stamp milling and amalgamation. *Mining Magazine* (London). Reprint, San Francisco: Mining and Scientific Press.

MacIntyre, Dan. 1995 (October). Telephone conversation with the author.

Marchant, Ward. 1993 (14 June). Summitville mine's tarnished legacy. *Fresno Bee*.

Matthes, François E. 1950. *The incomparable valley*. Berkeley: University of California Press.

Mazanec, Jane. 1993 (11 June). Failed Colorado mine a reform "poster child." *USA Today*.

McDonald, Douglas. 1988. *Bodie: Boom town—gold town!* Las Vegas: Nevada Publications.

McGrath, Roger D. 1984. *Gunfighters, highwaymen and vigilantes*. Berkeley: University of California Press.

McKowen, Ken. 1989 (Summer). Bodie. *California Parklands: The State Parks Magazine*, 3(3): 12–17.

McPhee, John. 1993. *Assembling California*. New York: Noonday Press.

Merrell, Bill. 1996. Personal notes. Merrell Collection, Bodie SHP, Sierra District, California Department of Parks and Recreation.

Merrell, Brownell. 1991. Map of Bodie, Bodie, Mono County, CA, (in) 1880. Author.

Mono-Alpine Chronicle. See *Bridgeport Chronicle-Union.*

Mono County Records. 1861–1918. Vols. 1–4, miscellaneous records; Vol. 5, Superior Court record, Bodie murder trials 1891 and 1899–1900; Vol. 6, Bodie court records 1907–1917; Vol. 7, miscellaneous Mono County court records and business license applications 1870–1889 (BANC MSS C-A 276). Bancroft Library, University of California at Berkeley. Manuscript.

———. 1862. Records of Bodey Mining District, CA: Books A and B. Mono County Courthouse, Bridgeport, CA.

———, *Great Register of Mono County.* 1860–1880, 1907–1917, 1934, 1935. Mono County Courthouse, Bridgeport, CA.

———. 1872–1884 (BANC xfF868.M6.M5). Bancroft Library, University of California at Berkeley. Manuscript.

———, *Miscellaneous birth records.* Mono County Courthouse, Bridgeport, CA.

———, *Miscellaneous marriage records.* Mono County Courthouse, Bridgeport, CA.

———, *Records of Bodey Mining District, CA: Book L, Mining deeds.* 1874. *Apportionment,* Vol. 3. 1865. Mono County Courthouse, Bridgeport, CA.

Moore, Thomas. 1969. *Bodie ghost town.* Cranbury, NJ: A. S. Barnes and Company.

Morgan, Jon Davis. 1995. The United States. *Mining Annual Review—1995: The Mining Journal and Mining Magazine,* pp. 63–69.

Morgan, Oscar. n.d. Personal reminiscences. Unit History Collection, Bodie SHP, Sierra District, California Department of Parks and Recreation. Photocopy.

Murray, Stewart. 1995. Precious metals and minerals: Gold. *Metals and Minerals Annual Review—1995: The Mining Journal and Mining Magazine,* pp. 15–19.

Nadeau, Remi. 1965. *Ghost towns and mining camps of California.* Los Angeles: Ward Ritchie Press.

Nevada Department of Wildlife (NDOW) Mining Program. 1999. Retrieved from the Web at http://dcnr.nv.gov/nrp01/res95.html. Accessed in 2002.

Nielson, Carl. 1989 (19 October). Galactic ghostbusters: The threat to Bodie. *Mammoth Times.*

Nutter, Allen. 1977 (13 December). Letter to Bodie SHP. Unit History Collection, Bodie SHP, Sierra District, California Department of Parks and Recreation.

Obmascik, Mark. 1993a (21 February). Mine disaster worsens to tune of $33,000 a day. *Sunday Denver Post.*

———. 1993b (2 May). Summitville: A symbol of mine abuse. *Sunday Denver Post.*

———. 1994 (25 January). Summitville mess probed. *Denver Post.*

O'Neil, J. R., Silberman, M. L., Fabbi, B. P., and Chesterman, C. W. 1973. Stable isotope and chemical relations during mineralization in the Bodie Mining District, Mono County, California. *Economic Geology,* 68: 765–784.

Parr, J. F. 1928 (May). Reminiscences of the Bodie strike. *Yosemite Nature Notes,* 7: 33–38.

Peak, Ann. 1975. *Cultural resource assessment of the Bodie Road from State Highway 395 to a point near the State Park of Bodie.* Mono County: Department of Public Works. Sacramento: California State Library. Photocopy.

Peterson, Gunnar "Pete." 1986 (30 July). Interview by Ranger Jack Shipley. Bodie, CA. Oral History Collection, Bodie SHP, Sierra District, California Department of Parks and Recreation. Tape recording.

Piper, William. 1935. Interview with unnamed anthropologist. Oral History Collection, Bancroft Library, University of California. Microfilm.

Poole, William. 1989 (2 July). Showdown at Bodie. *San Francisco Chronicle.*

Powers, Stephen. [1877] 1976. *Tribes of California.* Berkeley and Los Angeles: University of California Press.

Press, Frank, and Siever, Raymond. 1974. *Earth.* San Francisco: W. H. Freeman and Company.

Pupich, Mark. 1995 (August). Conversation with the author. Bodie, CA.

Quigley, Lesley, and Anderson, Edward. n.d. *Bodie—the Klipstein-Rosecrans years, 1935-1942.* Unit History Collection, Bodie SHP, Sierra District, California Department of Parks and Recreation. Photocopy.

Reiger, Ted. 1991 (July-August). Saving history's gold in the Old West. *Historic Preservation News,* pp. 1-2, 20.

Rensch, Hero Eugene, and Rensch, Ethel Grace. 1932 and 1933. *Historic spots in California.* Palo Alto, CA: Stanford University Press.

Rickard, T. A. 1922. *Mining and Scientific Press.* Interviews with mining engineers, Tom Leggett. San Francisco: Mining and Scientific Press.

Robson, John H. "Jack." 1994 (November). Interview by the author. Bodie, CA.

Rosen, Ruth. 1982. *The lost sisterhood: Prostitution in America, 1900-1918.* Baltimore: Johns Hopkins University Press.

Russell, Carl P. 1927 (31 December). Bodie, dead city of Mono. *Yosemite Nature Notes,* 6(12): 89-96.

———. 1928 (February). Early mining excitements east of Yosemite. *Sierra Club Bulletin,* 13: 40-53.

———. 1929 (November). The Bodie that was. *Touring Topics,* 21: 14-20.

Russell, Israel C. [1889] 1984. *Quaternary history of the Mono Valley, California.* Eighth Annual Report of the USGS [U.S. Geological Survey], 1889. Lee Vining, CA: Artemisia Press.

Salmon, Mary H. n.d. *Bodie—the other life.* Unit History Collection, Bodie SHP, Sierra District, California Department of Parks and Recreation. Manuscript.

Sartor, Louise. 1952 (December). I remember Bodie. *Desert Magazine,* 15: 12-15.

Save Bodie! Committee. 1990. *The Bodie experience—worth more than gold.* Sacramento: California State Library.

Scanavino, Steve. 1968 (3 June). Letter to Supt. Robert Freznel. Unit History Collection, Bodie SHP, Sierra District, California Department of Parks and Recreation.

Seagraves, Anne. 1994. *Soiled doves: Prostitution in the early West.* Hayden, ID: Wesanne Publications.

Shipley, Bodie Jack. 1995 (March and August). Conversation with the author. Bodie, CA.

Silberman, Miles. 1997. Telephone conversation with the author.

Smith, Gordon (Bodie). 1996. Exchange of letters with the author, from Bath, England.

Smith, Grant H. *Bodie*. 1925. *Bodie, the last of the old time mining camps* (BANC F868. M6S59). San Francisco: California Historical Society. Bancroft Library, University of California at Berkeley. Manuscript.

Smith, Herbert. 1934. The Bodie era. Sacramento: California State Library. Typescript.

———. n.d. Notes and pictures on western mining (BANC MSS C-R 31). Bancroft Library, University of California at Berkeley. Typescript.

Smith, Silas B. 1935. Interviews with unnamed anthropologists. Oral History Collection, Bancroft Library, University of California. Microfilm.

Sprague, Dr. Robert W. 1995 (July), 1996 (July). Conversations with the author. Palos Verdes, CA, and Berkeley, CA, respectively.

Standard. Also known as *Bodie Daily Standard, Daily Bodie Standard, Bodie Standard, Bodie Weekly Standard,* and *Weekly Standard-News*.

Standard Consolidated Mining Company Annual Report. 1879, 1880 (BANC xF862.3.S7). Bancroft Library, University of California at Berkeley. Manuscript.

State of California. 1850, 1860, 1910. California State Census. Sacramento: California State Library.

State of California, Department of Parks and Recreation. 1979. *Bodie State Historic Park resource management plan, general development plan and environmental impact report*. Sacramento, CA.

———. 1989. *Visitor survey*. Sacramento, CA.

———, Resources Agency. 1931 (January). Claim Map, Bodie Mining District, Mono County, CA. Surveyed and compiled by D. W. Ormsbee, G. E. Sacramento.

———. 1979 (March). Mine Workings of Bodie Bluff–Standard Hill Area, Bodie Mining District, Mono County, CA. Revised by Charles Chesterman. Sacramento.

———. 1991 (March). Bodie State Historic Park, Land Ownership. Map. Sacramento.

———. 1996 (December). Bodie State Historic Park, land ownership. Retrieved from the Web at F\:DATA\SPATIAL\685\324\ACAD\BODIE_SHP.DWG. Accessed in 2001. Map. Sacramento.

———. 1997. Travel news: Bodie State Historic Park now twice as big, thanks to help from Caltrans and Wildlife Conservation Board . . . Contact: Brad Sturdivant, Supervising Ranger, Bodie SHP.

———, Bodie State Historic Park. 1962–1996. Miscellaneous park records.

State of California, Division of Mines, Walter Bradley, State Mineralogist. 1934. *California Journal of Mines and Geology: Quarterly Report of State Mineralogist*, 30(1). Sacramento.

State of Colorado, Colorado Department of Public Health and Environment, Hazardous Materials and Waste Management Division. 1999. Remedial programs: Summitville mine. Retrieved from the Web at http://www.state.co.us/gov_dir/cdphe_dir/hm/rpsummit.html. Accessed in 1999.

State of New York. 1840, 1850, 1860. New York State Census. Church of the Latter-day Saints Library, Oakland, CA.

Tagliahue & Garrard. 1860s. Map of Bodie. San Francisco: Tagliahue & Garrard.

Tracy, Marylin "Fern" Gray. 1994 (December). Interview by the author. Oakland, CA.

U.S. Congress, House of Representatives. 1994 (4 October). *California Desert Protection Act of 1994, Conference Report, Title X: Bodie Protection Act of 1994.* 103rd Congress, 2nd session, 1994. Report 103-832.

U.S. Department of Health and Human Services, U.S. Public Health Service, Agency for Toxic Substances and Disease Registry, Division of Health Assessment and Consultation. 1997. Public Health Assessment, Summitville Mine, Del Norte, Rio Grande County, Colorado, Cerclis #COD983778432. Retrieved from the Web at http://www.atsdr.cdc.gov/hac/pha/summit/sum_toc.html. Accessed on 5 July 2002.

U.S. Department of the Interior, Bureau of Land Management. 1987. *Wilderness recommendations, Benton-Owens Valley-Bodie-Coleville study areas, final environmental impact statement.* Washington, DC.

———. 1998. Press release CA-99-02: Entire ghost town of Bodie now publicly owned. Contact: John Dearing, Steve Addington. Washington, DC.

———. 1990a (25 April). Bodie Historic District, National Historic Landmark, Mono County, CA. NPS Draft, DNRP.WRO (included in *Fiscal year 1990 report to Congress on threatened and damaged national historic landmarks*). Washington, DC.

———. 1990b. Press release regarding Bodie Historic District (included in *Fiscal year 1990 report to Congress on threatened and damaged national historic landmarks*). Washington, DC: Author.

U.S. Federal Census. 1850-1950. Sacramento: California State Library.

Van Loan, Charles. 1915 (18, 25 September). Ghost cities of the West: Bad, b-a-d Bodie. *Saturday Evening Post,* pp. 8-9 and 45-50; 18-19 and 55-58.

Voss, Marjorie Dolan Bell. 1994 (December). Interview by the author. Reno, NV.

War Powers Act of 9 March 1933. Retrieved from the Web at http://chansen.tzo.com/Subjects/Scams/Articles/WarPowersAct.html. Accessed on 4 July 2002.

Warren, Mrs. Ken. 1988 (29 September). Conversation with Bodie SHP staff member. Oral History Collection, Bodie SHP, Sierra District, California Department of Parks and Recreation.

Wasson, Joseph. 1878. *Bodie and Esmeralda.* San Francisco: Spaulding, Barto, and Company, Steam Book and Job Printers.

Wedertz, Frank S. 1969. *Bodie, 1859-1900.* Bishop, CA: Chalfant Press.

Weekly Standard-News. Also known as *Bodie Daily Standard, Bodie Standard, Daily Bodie Standard, Standard,* and *Bodie Weekly Standard.*

Whitehead, Mark [geologist]. 1996. Telephone conversation with the author.

Wiggers, Raymond. 1993. *The amateur geologist.* New York: Franklin Watts.

Williams, George, III. 1981. *The guide to Bodie and eastern Sierra historic sites.* Dayton, NV: Tree by the River Publishing.

Young, Otis E., Jr. 1970. *Western mining.* Norman: University of Oklahoma Press.

Young, Rick, and Noyes, Dan. 1994 (14 August). The road to Summitville, a gold mining debacle. *New York Times.*

Yung, Dr. Judy. 1996 (15 May). Telephone conversation with the author.

INDEX

Note: Italic page numbers refer to illustrations.

Benton, 113

Benton Mono Weekly Messenger, 28

Billeb, Dolly, 168

Billeb, Emil, 121, 123, 165, 168, 202n4

Bishop Creek Vigilance Committee, 56

Bishop Paiute Shoshone Cultural Center,
     192n5, 193n7

Blackhawk Mine, 114

Blackwood, Dr., 96, 113

Blake, William P., 13

blankets, 192

blasting, 39, 40, 173

boardinghouses, 16-17, 18, 76, 127, 152. *See
     also* brothels

Bodey, Wakeman L., 196, 197-98

"Bodey Mining District," 5

Bodie: abandoned property in, 116, 165;
     automobiles in, 120-22; "bad man"
     reputation of, 24-25, 57; benefit func-
     tions in, 54; boardinghouses in, 16-17,
     18, 76, 127, 152; Bonanza Street section,
     58; J. Ross Browne on traveling to, in
     1868, 15; businesses in, 51-52; caretakers
     of, 162-65; Chinatown, 58-59, 65, 97,
     108, 110, 114; Chinese immigrants and,
     99, 113; Chinese rail-worker crisis,
     105-6; churches and houses of worship
     in, 97, 114, 124; "circulating library," 53;
     condition of streets in, 51, 104; crime
     in (*see* crime); curfews, 115; current
     state of, 1-2, 175; decline of, 104, 106-8,
     110-16; demand for wood in, 26 (*see also*
     firewood; wood-and-lumber business);
     1877 development into a town, 19-21;
     difficulties in leaving, 131-32; early
     history of, 14-15; emigration from, 98;
     entertainment in (*see* entertainment);
     ethnic clusters in, 59; fires in, 22,
     140-43, 202n7; first inhabitants of, 187;
     flight of people in debt, 111; gamblers
     and swindlers in, 23; gangsters and, 137,
     140; geology of, 2-3, 177-80; as a ghost
     town, 162-65; ghost town reputation
     of, 132, 133; growth of, 22-24, 25-26,
     27-29; guns and gunfights in, 45, 57-58,
     59, 124; holidays in (*see* holidays);
     "hoodlums" in, 112; hotels in, 49, 50,
     76; hydroelectric power and, 116-17;
     jail house, 97; Kuzedika inhabitants

(*see* Kuzedika); lack of drainage and
     sewers in, 51; law enforcement in, 20-
     21, 97; life during the Great Depression,
     144-60; life in the early 1900s, 119-31;
     life in the final years, 133-40; life in the
     golden years of, 48-58; Main Street, *2,*
     49, 55; the mysterious Dr. Blackwood,
     96, 113; newspapers in, 53-54; nightlife
     in, 63-66; opium dens in, 59, 65,
     67-68; original name as "Bodey,"
     199n1; plumbing in, 148; in pop cul-
     ture, *171;* population in (*see* Bodie popu-
     lation); possible founding fathers, 196-
     98; post office, 19, 159, 162; prehistoric
     history of, 3; presence of stamp mills
     and mine workers in, 51; Prohibition
     in, 135, 136; "Railroad Station," 105;
     restaurants in, 18, 22, 49-50, 126-27;
     saloons and saloon life in, 57, 63-64,
     65-66, 108, 126-27; schools and
     schooling in, 20, 28, 52, 76, 148-51;
     Benjamin Silliman on conditions for
     the development of, 14; smells of,
     50; social organizations and secret
     societies in, 60-63; social structure of
     (*see* Bodie society); stagecoaches to, 49,
     122; standard of conduct in, 59-60,
     112; stock decline in 1881 and, 99,
     104; Sunday closure law, 108; temp-
     erance lecturers, 108; time line of,
     7-9; traveling psychics, 106; Treloar-
     DeRoche affair, 100-104; violence in,
     24-25, 29, 96-97; water of, 50; whiskey
     of, *56,* 66, *67,* 135; winters in (*see* win-
     ters); women in business, 52

Bodie, Elizabeth, 197

Bodie, W. S.: Terrence Brodigan and, 71;
     death of, 5; discovery of gold by, 4-5,
     39; eulogy for, 30; mystery of the
     identity of, 5-6; possible identities of,
     196-98; remains of discovered and
     reburied, 29-30

Bodie, William Smith, 6, 197-98

Bodie and Benton Railway, 113

Bodie and Benton Railway and Commercial
     Company, 110

Bodie and Esmeralda (Wasson), 22-23

"Bodie Bill," 140-41

Bodie Bluff, 3, 15, 18, 23, 96, 178

dwellings: building materials, *16;* burglary and vandalism of abandoned homes, 165; in early Bodie, 15; miners' cabins, 12; rumor of people abandoning their possessions in, 134

electricity, 116-17, 127
"elephant," 51
Empire Company of New York, 13-14, 15, 16-17
"enrichers," 180
entertainment: athletic competitions, 52, *125;* dances and balls, 53, 158-59, 164; held in the Miners' Union Hall, 52-53, 158-59
environmentalists, 172
Environmental Protection Agency (EPA), 172, 173
Ephydra hians. See kutsavi
Esmeralda Mining District, 10
Esmeralda Union, 16
Essington, Peter, 17

Fair, Ellen, 81-82
farms, 15
Farnsworth (police officer), 100-101
faults, 180
Featherstone, Ken, 124
federal census, 95
Ferguson, Rev. R. D., 30
"Fire in the head," 66
"Fire in the hole!," 39
fires: in Bodie, 22, 140-43, 202n7; destruction of the Syndicate mill and cyanide plant, 162
firewood: availability in the winter of 1881-1882, 111-12; demand for, 26; difficulties in the winter of 1879-1880, 95; to fuel steam engines, 18-19; hauled by mules, 27, *28;* for heating homes, 15; problems of theft, 27; sources of, 26-27; winter shortage in the early 1900s, 123
fissure veins, 180
flat claims, 10
Floyd, Pretty Boy, 137, 140
flu epidemic, 145
flux, 37, 200n5
foot binding, 89-90
footraces, *125*

Fortuna vein, 180
"40 mesh" screens, 183
'49er Argonauts, 3-4
Fourth of July celebrations, 28, 55-56, 107, 113, 124-25
"Four Virtues," 91
Fulton, Charlie, 151, 153
funerals, 124; of the Chinese, 89

Galactic Resources Ltd., 172, 173, 174, 186, 202n5, 203n3
gamblers, 23, 78-79
gangsters, 137, 140
Garfield, James A., 30, 107
gas signs, *121*
General Electric, 116
Genoa (Nev.), 19, 147
geology: of the Bodie Mining District, 177-80; doming, 178; fault formation, 180; formation of gold veins, 179-80; plugging of volcanic vents, 179, 180
ghost stories, 129-30, 156
ghost towns: Bodie as, 162-65; Rhyolite, 173
Gilson and Barber store, 71
Godward, Billy "Bodie Bill," 140-41
Goff, John, 45, 46
gold: consumer demand for, 171; discovery in the Bodie Mining District, 4-5; discovery in the eastern Sierra, 3-4; extracting from ore, 42-43, 181-86; geology of, 3; industries used in, 171
gold bullion: gold and silver mixed in, 183; mailing of, 39; percent gold in, 39; production in 1881, 107, 110, 111 (*see also* gold production); shipped in 1880, 48
gold extraction techniques: arrastras, *11,* 42, 181-82; ball mills, 183-84; cyanide heap-leach method, 184-85; cyanide process, 184; *De Re Metallica* and, 181; ore rockers, 182; stamp mills, 183. *See also individual techniques*
gold mines and mining companies: J. Ross Browne's account of being inside, 33, 36, 40; Marian Bryant's description of, 39; closings in 1882, 114-15; consumer demand for gold and, 171; decline in 1880, 98; demand for wood in, 26; Pat Dolan Goodwin's description of, 160; halting of work in 1881, 104; hoisting

in, 31–32, 34–36; the Jupiter/Owyhee
dispute, 45–47; operational in 1906,
125; pumping water from, 107; stock
certificates, 51; switch to electric power
in, 117; tax assessments and, 23, 114;
"timbering," 26; wages paid by, 51. *See
also* Bodie Mining District; *individual
mines and mining companies*
gold mining: blasting, 39, 40; corporate,
11–14, 17; "deep mining," 104; extracting
gold from ore (*see* gold extraction techniques); of fissure veins, 180; hazards
of, 41–42; hoisting, 31–32, 34–36;
methods for retrieving ore, 39, 41;
objections to, continuing at the current
time, 172–73; open-pit, 140, 172; pattern
of discovering and selling, 11; the War
Powers Act of 1933 and, 202n8
gold ore: assaying, 37–39; deposition of,
178; extracting gold from (*see* gold
extraction techniques); formation of
veins, 179–80
gold production: in 1877, 21; in 1878, 23; in
1879, 25; in 1881, 107, 110, 111; in 1882
and 1883, 115; impact of the cyanide
process on, 184
gold seekers: in 1848 and 1849, 3–4
gold shipments: after 1912, 132; by mail, 39
gold veins: formation of, 179–80; mining
of, in the Bodie Mining District, 180
Goodshaw claim, 39
Goodshaw Mine, 104
Goodwin, Ed: on burglary of abandoned
property, 165; on gangsters in Bodie,
137; memories of life in Bodie, 148, 158,
159–61
Goodwin, Pat Dolan: on gangsters in Bodie,
137, 140; on J. S. Cain, 144; memories
of life in Bodie, 119–20, 158–61; on the
noise of the stamp mill, 145
grave-site fences, *166*
Gray, Ed, 129, 145, 146
Gray, Larry, 129
Gray, Lauretta Miller, *128;* biographical
sketch of, 129; departure from Bodie,
130; on Depression-era life in Bodie,
145; on Fourth of July celebrations,
124–25; in a Fourth of July footrace,
125; on the Great Fire of 1932, 143; on

Kuzedika living in town, 85; on Rosa
May, 80; on Prohibition in Bodie, 135;
on saloons in the early 1900s, 126; on
winters in Bodie, 123, 153
Gray family house, 169–70
Great Depression: life in Bodie during,
144–60
Great Register of Mono County, 19
"Great Storm" of 1879, 29
Great Western mining claim, 16
Green Creek power plant, 116–17
Green Street, 105, 112
Green Street schoolhouse, 20
Gregory, Spence, 162–63, 164
grizzly, 183
gulch claims, 10
gunmen, 57, 126
Gunn's saloon, 63–64
guns and gunfights: Howard Ball on, 124;
the Bodie caretakers and, 163, 164;
frequency in Bodie, 57–58, 59; in the
Jupiter/Owyhee dispute, 45; miners
and, 139

Harber, George, 200n9
Harper's Monthly, 15
Harrington, John, 81
Harvey, Clyde and Annis, 153
Hastings, H. F., 20
Hawthorne (Nev.), 96, 116, 127
headstones, 135, 165, *166*
heap-leach pad, 184
Heizer, Robert, 195
"high grading," 202n1
high school, 149
hill claims, 10
hoisting, 31–32, 34–36
hoisting engineers: Bob Bell on, 31–32, 34–
36; importance of, 31; strikes by, 43–45
hoisting works, *34*
holidays: Christmas, 125, 151; Fourth of July
celebrations, 28, 55–56, 107, 113, 124–25;
Thanksgiving, 110
Holt, Thomas, 112
Home Comfort stove, *174*
Homestake Mining Company, 140, 172,
202n5, 203n3
homicide rate, 57, 58, 200n3
homicides: frequency in Bodie, 57, 58;

in self-defense, 60; Treloar-DeRoche
affair, 100–104
Honey Lake Paiute Indians, 199n1
"hoodlums," 112
Hoover, Herbert, 125
Hoover, Theodore, 125
Horner, Mrs. Robert (née Marietta Butler),
14
Horribles, 55–56
horses: in winter, *122*
horse snowshoes, 156
Hotel de Kirgan, 97
hotels, 49, 50, 76
Hunnewill, N. B., 26
hunting: by the Kuzedika, 191
"hurdy-gurdy" girls, 77
Hydro Building, 147, 165
hydroelectric power, 116–17. *See also* Jordan
power plant
hydrogen cyanide, 185–86
hydrothermal fluids, 179

igneous rock, 177, 178, 179
incense sticks, 201n16
indentured servitude, 91
Independence Day. *See* Fourth of July
celebrations
Indian doctors, 193
Indian reservations, 195
Indians. *See* Kuzedika; Native Americans
Indian schools, 195
industrial accidents, 185–86
Irwin, Charlie, 55
Irwin, William, 20
Isabella Mine, 13
Isabella tunnel, 12, 199n1

J. S. Cain Company, 160, 162, 168, 172
jail house, 97
Janin, Alexis, 117
"jaw crushers," 199n1
Jeffrey pine, 188, 189
Jeffrey pine caterpillars. *See* piaghi
Jenny Lind restaurant, 18
Jewish residents, 82
Jim, Susie, 188, 189–90, 191, 192, 193
Johl, Eli, 76, 79–80
Johl, Lottie, 76, 79–80
"Johns," 201n12

Jones splitters, 37, 199n1
Jordan power plant: avalanche and rescue
at, 127, 130–31; construction of, 117, 127
Joseph, Dorothy, 162, 163–64
"joss sticks," 201n16
joss temples, 97, 201n16
Jupiter Mine, 45–47, 200n9
Jupiter/Owyhee dispute, 45–47
Jurassic period, 177

Kearnahan, Olsen & Co., 16
Keely, C. C., 133–34
king mold, 37
King Street, 58–59
King Street restaurant, 22
Kirgan, J. F., 20–21, 28, 97, 104
Kittrell, John, 101
Klipstein, Henry, 148, 160, 161
Klipstein, Virginia, 148
Kriel, Zady, 153
kutsavi, 3, 85, 188–89
Kuyui-dika, 199n1
Kuzedika: in Bodie, 59, 83–86; the Bodie
railroad and, 108; crafts of, 191–92;
description of, 3; diet of, 187–88;
hunting by, 191; impact of European-
Americans on, 193–95; kutsavi and, 3,
85, 188–89; medicine men, 193; as the
Mono Paiute Indians, 199n1; piaghi
harvest, 188, 189–90; pine nut harvest,
190–91; round houses of, 192–93;
spring onion harvest, 188; tribal social
structure, 193

Lafee, Helen, 59, 66
LaGrange, O. H., 20
laundry stoves, *88*
law enforcement, 20–21, 97
lawyers, 73–74
lead button, 37
lead oxide, 37
"leasers," 132
Lee, Toy, 92
Leeds, Esther, 85, 92
Lee Vining (Calif.), 76, 131, 136
Lee Vining Canyon, 4
Leggett, Tom, 116, 117–18
Leon, Sam, 136–37
Lewis, Arena Bell: Gordon Bell and, 147; on

Ella Cain's dog, 149; on Depression-era
life in Bodie, 145, 151; on Billy Godward,
141; magpie of, 150; on Prohibition in
Bodie, 135
libraries, 53
light bulbs, 127
Lincoln, Abraham, 71
Lincoln, Sarah, 71
liquor. *See* alcohol
liquor licenses, 14
"litharge," 37
Lockberg, Louis, 17
Lonahan, Johanna. *See* Treloar, Johanna
Lonahan
Loose, Ed, 17, 19
Loose, Warren, 19, 123
Loose, William, 19
"lotus" feet, 90
Lucky Boy, 108
lumber companies, 26, 73, 96, 97-98.
See also Bodie Railway and Lumber
Company
lynchings, 100-104

magma, 178
magma chambers, 178, 180
magpies, 147, 150
mail: shipping gold bullion by, 39; during
the winter, 151, 153
Mammoth Mountain, 177
*Manufacturer and Builder, a Practical Journal
of Industrial Progress,* 182
manway, 35
Mapstead, Diana, 131, 163, 164
"marginal ore," 162
marriage: California's Anti-Miscegenation
Law, 91-92
Masonic, 71, 108
Masonic Springs, 157
Matthew Vassar, 198
May, Rosa, 80, 127
Maybelle Mine, 114
McCallum, Mike, 114
McClinton, J. Giles, 29
McCone, A. J., 118
McDonald, Barney, 200n9
McDonald, Joe, 200n9
McDonnel, Alice, 159
McDonnel House, 157, 158, 159

McMillan House, 153
"mechanical shovels," 160
Mechanics' Union strike, 43-45
medicine men, 193
mercury, 42, *43,* 182, 183
Merrill, Charles, 117
Merrill-Krull process, 161
metamorphic rock, 177, 178
Methodists, 114, 124
Metzger House, 150
Mexican residents, 82, 83
Mill Creek, 95
Miller, Annie, *50, 126;* biographical sketch
of, 76; boardinghouse of, 76, 127, 143; at
Bodie dances, 158; the Great Fire of 1932
and, 143; opens a "circulating library"
in Bodie, 53; owner of the Occidental
Hotel, 76, 127
Miller, Lauretta. *See* Gray, Lauretta Miller
Miller, "Tuffy," 150
Miller, William, 76, 129
mill workers: wages, 51
mine cages, 32, 34-35, 36
mineralization, 179, 180
miners, *32, 41, 46;* assayers and assaying,
37-39; 1880 federal census and, 95; guns
and, 139; hazards of mining work and,
41-42; heroics of, 41-42; "high grading,"
202n1; hoisting engineers, 31-32, 34-36;
"leasers," 132; Mechanics' Union strike,
43-45; physical examinations and, 41;
precautions against marauders, 138-39;
prostitutes and, 78; responsibility for
gold shipments after 1912, 132; wages,
44, 51; working shifts, 43-44
miners' cabins, 12
Miners' Union: exclusion of Chinese people
from, 88; flag of, *44;* formation of, 21;
the Jupiter/Owyhee dispute and, 45-47;
the Mechanics' Union strike and, 44-45
Miners' Union, The (newspaper), 163-64
Miners' Union Hall, *7;* construction of, 21;
entertainments held in, 52-53, 158-59;
museum in, 170; religious services held
in, 22
mines and mining companies. *See* gold
mines and mining companies
mine shafts, 35
Mining Camp Days (Billeb), 121, 202n4